COMPARATIVE PLANETOLOGY

CONTRIBUTORS

DUWAYNE M. ANDERSON
BONNIE J. BERDAHL
K. BIEMANN
J. E. BILLER
ROBIN BRETT
FREDERICK S. BROWN
GLENN C. CARLE
MARGARET O. DAYHOFF
A. V. DIAZ
FAROUK EL-BAZ
ANNA-STINA EDHORN
D. FLORY
HERBERT FREY
GEORGE L. HOBBY
THOMAS C. HOERING
NORMAN H. HOROWITZ
D. W. HOWARTH
JERRY S. HUBBARD
RICHARD D. JOHNSON
HAROLD P. KLEIN
A. L. LaFLEUR
JOSHUA LEDERBERG
JOEL S. LEVINE
PAUL D. LOWMAN, JR.
A. O. NIER
A. I. OPARIN
L. E. ORGEL
J. ORO
T. OWEN
VANCE I. OYAMA
ALEXANDER RICH
D. R. RUSHNECK
ROBERT M. SCHWARTZ
P. G. SIMMONDS
PATRICIA A. STRAAT
JAMES C. G. WALKER

Comparative Planetology

Edited by

Cyril Ponnamperuma

Laboratory of Chemical Evolution
Department of Chemistry
University of Maryland
College Park, Maryland

ACADEMIC PRESS   **New York   San Francisco   London**   **1978**
A Subsidiary of Harcourt Brace Jovanovich, Publishers

ACADEMIC PRESS RAPID MANUSCRIPT REPRODUCTION

Proceedings of the Third College Park Colloquium
on Chemical Evolution — Comparative Planetology
Held September 29 — October 1, 1976 at the University of Maryland.

ACADEMIC PRESS, INC.
111 Fifth Avenue, New York, New York 10003

United Kingdom Edition published by
ACADEMIC PRESS, INC. (LONDON) LTD.
24/28 Oval Road, London NW1 7DX

LIBRARY OF CONGRESS CATALOG CARD NUMBER:

ISBN 0-12-561340-7

PRINTED IN THE UNITED STATES OF AMERICA

CONTENTS

LIST OF CONTRIBUTORS

Numbers in parentheses indicate the pages on which authors' contributions begin.

Duwayne M. Anderson (197, 219), Division of Polar Programs, National Science Foundation, Washington, D.C.

Bonnie J. Berdahl (7), NASA Ames Research Center, Moffett Field, California

K. Biemann (183, 191, 213), Department of Chemistry, Massachusetts Institute of Technology, Cambridge, Massachusetts

J. E. Biller (183, 197), Department of Chemistry, Massachusetts Institute of Technology, Cambridge, Massachusetts

Robin Brett (27), U.S. Geological Survey, Reston, Virginia

Frederick S. Brown (7), TRW Systems, Redondo Beach, California

Glenn C. Carle (7), NASA Ames Research Center, Moffett Field, California

Margaret O. Dayhoff (225), National Biomedical Research Foundation, Georgetown University Medical Center, Washington, D.C.

A. V. Diaz (197), NASA Langley Research Center, Hampton, Virginia

Farouk El-Baz (103), National Air and Space Museum, Smithsonian Institution, Washington, D.C.

Anna-Stina Edhorn (257), Department of Geological Sciences, Brock University, St. Catherines, Ontario L2S 3A1, Canada

D. Flory (197), Spectrix Corporation, Houston, Texas

Herbert Frey (79), Department of Physics and Astronomy, University of Maryland, College Park, Maryland

George L. Hobby (7), California Institute of Technology, Pasadena, California

Thomas C. Hoering (243), Carnegie Institution of Washington, Geophysical Laboratory, Washington, D.C.

Norman H. Horowitz (7), California Institute of Technology, Pasadena, California

D. W. Howarth (183, 213), Guidance and Control Systems, Litton Industries, Woodland Hills, California

Jerry S. Hubbard (7), Department of Biology, Institute of Technology, Atlanta, Georgia

Richard D. Johnson (7), NASA Ames Research Center, Moffett Field, California

Harold P. Klein (7), NASA Ames Research Center, Moffett Field, California

A. L. LaFleur (183, 213), Department of Chemistry, Massachusetts Institute of Technology, Cambridge, Massachusetts

Joshua Lederberg (7), Department of Genetics, Stanford University, Stanford, California

Gilbert V. Levine (7), Biospheries, Incorporated, Rockville, Maryland

Joel S. Levine (165), Atmospheric Environmental Sciences Division, NASA Langley Research Center, Hampton, Virginia

Paul D. Lowman, Jr. (51), Geophysical Branch, Goddard Space Flight Center, Greenbelt, Maryland

A. O. Nier (197), School of Physics and Astronomy, University of Minnesota, Minneapolis, Minnesota

A. I. Oparin (1), A. N. Bach Institute of Biochemistry, Leninsky pr, 33, Moscow B-71, USSR

L. E. Orgel (197), Salk Institute for Biological Studies, San Diego, California

J. Oro (197), Department of Biophysical Sciences, University of Houston, Houston, Texas

T. Owen (183, 191, 213), Department of Earth and Space Sciences, State University of New York, Stony Brook, New York

Vance I. Oyama (7), NASA Ames Research Center, Moffett Field, California

Alexander Rich (7), Department of Biology, Massachusetts Institute of Technology, Cambridge, Massachusetts

D. R. Rushneck (197, 213), Interface, Inc., Fort Collins, Colorado

Robert M. Schwartz (225), National Biomedical Research Foundation, Georgetown University Medical Center, Washington, D.C.

P. G. Simmonds (197), Organic Geochemistry Unit, School of Chemistry, University of Bristol, Bristol, England

Patricia A. Straat (7), Biospheries, Incorporated, Rockville, Maryland

James C. G. Walker (141), National Astronomy and Ionosphere Center, Arecibo Observatory, P. O. Box 995, Arecibo, Puerto Rico

PREFACE

The latest information reaching us from a number of planetary missions demands the consideration of many aspects of the evolution of the solar system in a comparative fashion.

The Third College Park Colloquium on Chemical Evolution was held at a time when fresh data from Mars poured down to Earth in an almost unending stream. For the first time in human history, a survey of a planetary surface was in progress simultaneously for geology, chemistry, and biology. To highlight the implications of the data obtained from the Viking mission to Mars, the results were discussed with the backdrop of data available from the missions to other planets.

The interiors of planets, crustal evolution, the origin of planetary atmospheres, and the question of life itself provided the basis for the discussions among the astronomers, geologists, chemists, and biologists who were present at this meeting.

We are pleased to present this material to the growing body of scientists interested in the problem of the origin and evolution of life in our solar system.

1
THE NATURE AND ORIGIN
OF LIFE[1]

A. I. Oparin
USSR Academy of Sciences

The problem of the origin of life belongs to the most principal problem of natural science, since until this problem is
solved the human intellect cannot understand the nature of life
itself. Certainly, we may study and understand in a very deep
and comprehensive manner what substances, structures, and processes are forming the basis of the organization of modern organisms. Nevertheless, if we start with these principles alone,
we can never know why its organization is exactly as it is, in
particular, why it is so "purposeful", that is, why are the
structures of all integral parts of the organism (molecules,
organoid, organs) so perfectly adjusted to their functions and
why is the whole organism adapted to live under given environmental conditions?

At the turn of the twentieth century the problem of the origin of life was in a very critical situation. It was considered that in principle it could not be solved by methods of objective scientific study, being rather the field of faith than
of knowledge. There was no explanation for the appearance of
primitive organisms, the forerunners of all living beings on
the Earth. This situation berobbed the evolutionary system of
Darwinism of its fundamentals and caused deep disappointment
among natural scientists.

At that time one of the basic obstacles in solving the problem of the origin of life was the dominating belief, derived
from every experience, that in nature organic substances could
appear only biogenetically, i.e., via their synthesis by living
organisms. Certainly at that time organic chemists had synthesized in laboratories various complex organic compounds.

[1]Paper presented at the Special Session of the USSR Academy of Sciences on the occasion of its 250th anniversary.

This fact, however, did not alter the ideas stated previously, since it was believed that man, who is a living being, is capable of selecting the necessary sequence of reactions; this situation cannot arise in the inorganic world.

Nevertheless, it was absolutely impossible to conceive a sudden appearance of organisms, even the most primitive ones, directly from inorganic substances (carbon dioxide, water, nitrogen); this seemed to deadlock the problem of the origin of life.

In the early 1920's I formulated a concept that contradicted this generally accepted opinion. According to my hypothesis, the monopoly of the biogenic syntheses of organic substances is characteristic of the present epoch of the Earth's history.

In the early stages of its formation, when our planet was lifeless, Earth was the place of the abiogenic syntheses of carbonaceous compounds.

A prebiotic evolution of these compounds led to gradual accumulation of more and more complex substances that served as materials for the formation of individual phase-separated systems. As a result of natural selection, the latter could give rise to the appearance of brobionts and then to primitive living organisms.

During the last fifty years these concepts have received strong factual support from the studies of many scientists working in different countries and in different branches of science.

First of all, recent radioastronomical data indicate the presence in different regions of intersteller space of various carbonacaoeous compounds, sometimes rather complex and polyatomic ones. These molecules must be synthesized on the interstellar dust particles intensively irradiated with stellar radiations.

Thus, at present we have obtained direct evidence that the abiotic syntheses of organic compounds might have taken place not only before the emergence of organisms, but even before the formation of our planet.

We have good reason to conclude that the Earth has "inherited" from space a large quantity of *abiotic* organic compounds, first of all *nonvalative* organic ones to become a part of the initially more homogenous solid body of our planet.

Obviously, we can hardly hope to detect these "juvenile," in the strict sense of the term, cosmic organic substances in the sediments of the contemporary Earth's crust. A substantial portion of them should have undergone pyrolysis, the volatile products having reached the surface of the Earth and joined its atmosphere. Another portion could have acted, after some chemical modifications, as the material for the formation of probionts and primitive organisms.

After all, the third portion might be assimilated by the primitive organisms and became involved with their metabolism. Since the rate of metabolic transformations of organic material strongly exceeded that of primordial abiotic processes, it almost completely concealed the abiotic evolution of carbonaceous compounds in the Earth's crust. For this reason, the organic matter detected in the crust is usually considered to be of purely biological origin, formed as a result of decomposition of living matter.

Today it is quite obvious, however, that even the substances of biological origin that underwent deep degradation, even those that were completely mineralized, may again give rise to the complex organic molecules as a result of catalytic processes on the surface of mineral particles of the Earth's crust.

These reactions are of purely abiotic nature and could be illustrated by the Fisher-Tropsch reaction or by a similar process of amino acids formation from CO_2, H_2, and NH_3. In this light, how should we regard the nature of the substances formed and of the very process of their formation?

Certainly, their carbon atoms have already passed through living systems and in this respect they must be considered the substances of biological origin. However, the processes of their secondary synthesis are abiotic in nature and are similar to what had occured in the Earth's crust before life emerged. From this viewpoint, these processes are of great interest for the study of evolution of organic molecules that led to the origin of life.

Unfortunately, until quite recently this evolution has been investigated mainly in modeling experiments reproducing synthetic processes in the gas phase or in aqueous solution, in the atmosphere and hydrosphere of the primitive-Earth. The study of chemical evolution under natural conditions is strongly complicated by the powerful biological processes that occur in the Earth's crust. For this reason, no significant advance has been made yet in this field.

The majority of studies are connected with the analysis of decomposition of biogenetic organic substances and only a few attempts have been made to find conditions under which the abiotic syntheses might be studied with no influence of biological processes.

The most reliable information may be obtained at the study of nonterrestrial bodies: planets, comets and meteorites-carbonaceous chondrites. At present there is no doubt about the abiotic origin or organic material in meteorites.

Its formation has proceeded under conditions somewhat different from those of the primitive Earth: nevertheless, the specific character of the synthetic processes is quite obvious. We have a right to expect also that very important results will appear from the forthcoming exploration of Mars.

A rather good approximation of the conditions of the primitive Earth may be reached in the modeling experiments, still not numerous, reproducing more and more complex syntheses of organic

substances, which are adsorbed on the surface of some mineral
structures and there undergo various transformations catalyzed
by inorganic catalysts. It becomes clear that the characters
of both the adsorption process and the catalytic synthesis can
be very selective and specific. Even the simple Fisher-Tropsch
reaction results in products that, due to their structure, are
much more specific than the substances formed in free-radical
reactions or in the simple interaction of iron carbide with acid.

The whole complex of the aforementioned data permits us to
picture a possible pathway of abiotic formation of various car-
bonaoeous molecules, even very complex ones, in the local territories of the primitive Earth.

Today, the sharpest question among others intriguing the
scientists studying the origin of life is the problem of transition from chemical evolution to a biological type of organization
of matter.

This transition is characterized by the appearance of the
following new features, which in total are peculiar to living
systems only.

1. The ability to oppose an increase of entropy.
2. The appearance of purposeful organization of living beings,
i.e., adaptability of intra- and inframolecular structure of the
parts to fulfill their functions and of a whole organism to live
under the given environment.
3. The emergence of the characteristic life transfer of information heredity.

All these features could arise only at the formation of multimolecular phase-separated systems.

According to the second principle of thermodynamics, the inorganic world is characterized by an increase of entropy. On
the contrary, the organisms are capable of creating order out of
disorder, of opposing the increase of entropy, and even of decreasing it, while maintaining a high level of free energy.

This situation can take place only in so-called open systems,
the ones that are isolated from external environment by a distinct separation surface, but nevertheless are consuming matter
and energy from it.

The emergence of the phase-separated open systems was an
obligatory step for the formation of the other characteristic
of life: the purposeful organization of a whole organism and of
its parts.

This must appear on the basis of natural selection of the
whole systems, their metabolism, and their structure. According to the idea that has dominated scientific literature even
recently, the original formation of primitive living beings had
to be preceded by the appearance, in the local territories of
the Earth, of the protein and nucleic acid molecules whose
structures were internally organized and purposefully adjusted

to their functions. After this concept, the self-assembly of
these molecules could give rise to primitive organisms, as a
machine is assembled from prefabricated parts.

However, in this latter example, the details are purposefully
adjusted to the work of the whole machine, because they had been
produced after a prefabricated plan or a design.

When discussing problems of the origin of life, we cannot
assume the existence of a plan of creation. However, without
it, how could the protein or nucleic acid molecules appear whose
internal structure was well adapted to fulfill functions they
will carry out in the whole living system produced by their self-
assembly.

Such a concept is somewhat similar to the idea of Empedocles,
who alleged that originally the parts of the living body had to
appear (hands, feet, eyes, etc.) and later their aocretion brought
about the whole organism.

The functional adaptivity of the biologically significant
molecules could develop only in the process of evolution, as a
result of natural selection. We must realize that there were
neither the discrete nucleic acid molecules, nor even the protein
enzymes formed after their direction, which were subjected to the
selection, but the integral phase-separated systems consisting
of nonspecific polypeptides and polynucleotides.

Thus, the parts did not determine the purposeful organization
of the whole, but development of the whole brought about the
emergence of the purposeful adaptivity of the molecular structure
of the parts.

The natural selection of phase-separated systems has conditioned the emergence of the third feature of life:heredity. Originally, at the molecular level, only the protein- and nucleic
acid-like polymers could arise. Their association in the multimolecular phase-separated systems led to their interactions and
thus to the mutual coordination of their molecular structures and
biological functions as a result of natural selection of the integral open systems.

The ways of formation of such code relations may be understood through investigation of natural selection of the integral
protein-nucleic acid systems. Universality of the contemporary
genetic code is conditioned by the fact that at present any
change would be lethal, thus being a strongly antiselective factor.

Thus transition from a chemical evolution to a biological
one needed the formation of the discrete phase-separated open
systems capable of interacting with the surrounding environment
by utilizing its substances and energy for growth and to be a
subject of natural selection. How can we approach the objective
scientific study of this stage of evolution?

First of all it may be done in modeling experiments. Segregation of the phase-separated multimolecular systems from a

homogenous solution of organic substances often proceeds in
nature and we permanently observe it in the laboratory when
working with high molecular substances. Therefore, we cannot
only speculate about formation of these systems but even re-
produce, in the laboratory, the formation of various systems
that may serve as models for the structures having emerged at
the primitive Earth.

Among numerous possible models I have chosen for my experiments coacervate drops since these structures acquire very easily
the properties of the open systems.

The coacervate drops do not need any specific molecular
structure for their formation but arise after a simple mixing
of various polymers, both natural and synthetic.

The mixing brings about self-assembly of the polymer molecules to form multimolecular phase-separated structures; coacervate drops, isolated from the surrounding solution by a distinct boundary and yet capable taking up the substances from
it, as the open systems do.

In our experiments many substances, for example, energy-rich
compounds, when entering the drops undergo certain transformations, in particular polymerization. The polymers formed e.g.,
polyadenylate, become the components of the drops itself; at
their expense the drops grow, expanding their size and mass.
It is very important that the rate of their growth depends on
the composition and structure of any discrete drop. As a result
the different drops, when placed into the same solution, grow
with different velocities.

Thus, we have succeded in experimentaly demonstrating primitive natural selection, the principle that underlie all further
evolution of such systems on the way to the origin of life.

The attempt seems very promising to correlate our models of
phase-separated systems with the so-called organized elements
of the Orgeil meteorite or with organic bodies found in Archean
sediments.

2

THE VIKING BIOLOGICAL INVESTIGATION: PRELIMINARY RESULTS

Harold P. Klein, Vance I. Oyama, Bonnie J. Berdahl
Glenn C. Carle, Richard D. Johnson
NASA Ames Research Center, Moffett Field, California

George L. Hobby, Norman H. Horowitz
California Institute of Technology, Pasadena, California

Patricia A. Straat, Gilbert V. Levin
Biospheres, Incorporated, Rockville, Maryland

Joshua Lederberg
Department of Genetics, Stanford University, Stanford, California

Alexander Rich
Department of Biology, Massachusetts Institute of Technology
Cambridge, Massachusetts

Jerry S. Hubbard
Department of Biology, Institute of Technology, Atlanta, Georgia

Frederick S. Brown
TRW Systems, Redondo Beach, California

*Three different types of biological experiments on samples
of martian surface material ("soil") were conducted inside the
Viking lander. In the carbon assimilation or pyrolytic release
experiment. "CO_2 and CO were exposed to soil in the presence
of light. A small amount of gas was found to be converted into
organic material. Heat treatment of a duplicate sample prevented such conversion. In the gas exchange experiment, soil
was first humidified (exposed to water vapor) for 6 sols and
then wet with a complex aqueous solution of metabolites. The
gas above the soil was monitored by gas chromatography. A
substantial amount of O_2 was detected in the first chromatogram
taken 2.8 hr after humidification. Subsequent analyses revealed
that significant increases in CO_2 and only small changes in N_2
had also occurred. In the labeled release experiment, soil was
moistened with a solution containing several ^{14}C-labeled organic
compounds. A substantial evolution of radioactive gas was reg-
istered, but did not occur with a duplicate heat-treated sample.
Alternative chemical and biological interpretations are possible
for these preliminary data. The experiments are still in progress, and these results so far do not allow a decision regarding the existence of life on the planet Mars.*

We present here a preliminary progress report on the Viking
biological investigation, through its first month. Details
of the scientific concepts behind each of the experiments, as
well as examples of the kinds of results that are obtained when
these concepts are tested with the use of terrestrial samples,
have been described (Horowitz *et al.*, 1972; Levin, 1972; Oyama,
1972). The actual flight instrumentation and the tests to which
the flight instruments were subjected have also been described.
(Klein, 1974; Klein *et al.*, 1976).

During the manufacture of the flight instruments for the
biology experiments, rigorous clean-room techniques were employed to minimize airborne contamination (McDade, 1971) after
which the fully assembled flight hardware was heated at 120 ±
$1.7^{\circ}C$ for 54 hr in an atmosphere of dry 100% nitrogen prior to
shipment to the Kennedy Space Center. Here the instruments
were installed in the landers under clean-room conditions and

heated once more when the encapsulated landers were subjected
to terminal sterilization. This time the heating regime was
$112 \pm 1.8^{\circ}C$ for periods sufficient to reduce the spacecraft biological contamination loads to acceptable limits (Hall, 1975).

About a month after Viking I went into orbit around Mars,
the biology instrument was turned on briefly for the first time
since launch. At this time, 39 hr before separation, selected
valves within the instrument were automatically closed to prevent exhaust products from entering the instrument during the
descent phase when the instrument was powered down. On 22 July
1976, two days after landing, the instrument was again turned
on. With activation of both radioactivity detectors, background
counts were taken in dual- and single-channel counting modes.
A chromatogram was also taken, and the appropriate incubation
cells were rotated into position to receive surface samples.
The sample for the biology investigation reported here was acquired in the morning of sol 8 (a Mars day is called a sol and
equals 24 hr 39 min) from the surface at a depth of 0 to 4 cm
in an area consisting chiefly of fine-grained material. The
sample was introduced into the instrument via a soil processor
on top of the lander, which screened out coarse material, larger
than 1.5 mm: 7 cm^3 of the resulting smaller-grained material
was metered down into the biology instrument. Samples for the
individual biology experiments were metered and distributed
into the cells for subsequent use, as described below. The
temperature of the sample was below $0^{\circ}C$ during acquisition and
delivery, and was $9^{\circ}C$ during the period of storage in the test
cells prior to the initiation of the experiment. The major
events for the three experiments are outlined in Table I.

Our overall strategy called for relatively short incubation
periods for the first sample. If these proved negative, considerably longer periods could be used in later incubations.
Table II shows the various incubation sequences that are possible for the three experiments. The second Viking spacecraft
landed at a more northerly latitude and a colder environment.
After January 1977, at this site, incubation temperatures can
be significantly lowered within the biology instrument. Part
of the strategy, therefore, is to incubate martian soils at
these low temperatures. The first actual science data from the
biology instrument were returned from Mars four weeks before
this report was written. In this interval, during which the instrument functioned nominally, all three of the experiments

Table I Major Events Time Line for Biology Investigation

Earth date 1976	Mars time (sols) from landing	Pyrolytic release	Events during: Gas exchange experiment	Labeled release experiment
20 July 5:12 a.m. P.D.T.		Landing Initialize instrument		
22 July	2.98	Acquire soil		
28 July	8.29	Distribute soil		
	8.34		Seal test cell	
	8.36	Inject $^{14}CO_2$ and		
	8.39	^{14}CO;begin incubation		
29 July	8.60			Begin background count
	9.21		Add Kr, CO_2, He	
	9.22		Inject 0.5 ml of nutrient: begin incubation	
	9.33		Analyze gas	
	10.23		Analyze gas	Inject nutrient;begin incubation
	10.35		Analyze gas	
31 July	11.35		Analyze gas	
	11.4–13.4	Count background		
2 August	13.35	Terminate incubation		
	13.4–13.6	Pyrolyze: count first peak		
4 August	15.33		Analyze gas	
	15.8–16.3	Count background		
5 August	16.24		Inject 2.3 ml of nutrient	
	16.35		Analyze gas	
	17.0–18.0	Elute second peak and count		
	17.23			Inject nutrient
6 August	17.35		Analyze gas	
	18.35		Analyze gas	

Date	Time			
9 August	20.31		Analyze gas	Purge and dry test cell
	23.49			Begin background count
	23.59			Heat cleanup
	24.09			
13 August	24.52			Distribute soil to second test cell
	25.32		Analyze gas	
16 August	27.1–27.3	Sterilize second soil sample		
	27.4	Inject $^{14}CO_2$ and ^{14}CO; begin incubation		Begin background count
	27.46			
	28.21		Analyze gas	
18 August	28.22			Sterilize second sample
	29.24			Inject nutrient; begin incubation
	30.5–32.5	Count background		
	32.5–32.7	Terminate incubation;pyrolyze; count first peak		
	33.1–33.7	Count background		
	34.0–36.3	Elute second peak and count		
24 August	35.23			Inject nutrient
	36.28[a]			
	36.51[a]			
	37.53			Purge test cell
	37.64			Count background
27 August	38.14			Heat cleanup

[a] During the interval 36.28 to 36.51, power to the entire system was interrupted, according to prior arrangement.

TABLE II. *Viking Biological Investigation Sequences*

	Planned	Accomplished as of 27 August
Labeled release experiment		
Incubation of first sample (13.5 sols)[a]		X
Incubation of second sample (9.5 sols)		X
Extended incubation before conjunction (60 sols)[b]		
Possible "through-conjunction" incubation (100 sols)[b]		
Extended incubation after conjunction		
Cold incubation, post-conjunction[c]		
"Control" incubation if necessary		X
Gas exchange experiment		
Humid incubation (7 sols)		X
Wet incubation (30 sols)		X
Extended, wet incubation (85 sols)[b]		
Possible, wet incubation (85 sols)[b]		
"Control" incubation, if necessary		
Pyrolytic release experiment		
Incubation in the light, dry (5 sols)		X
Incubation in the light, wet (5 sols)		
Extended, dark incubation (35 sols)		
Possible "through-conjunction" incubation[b]		
Cold incubation, post-conjunction[c]		
"Control" incubation, if necessary		X

[a] *Sol, one martian day (24.6 hr).*

[b] *Possible only on Viking 1.*

[c] *Possible only on Viking 2, where incubation temperatures of around $266°K$ can be achieved.*

yielded data indicating that the surface material of Mars is chemically or biochemically quite active. Under normal circumstances, it would be premature to report biological experiments in progress before the data are amenable to ready interpretation. However, the unique nature of this investigation impels us to make this report, and we are fully cognizant of its preliminary nature.[1]

[1] *Additional accounts of each of the three biological experiments are being prepared by the principal investigators and their coinvestigators: N. H. Horowitz (pyrolytic release experiment), G. V. Levin (labeled release experiment), and V. I. Oyama (gas exchange experiment).*

THE CARBON ASSIMILATION EXPERIMENT

The pyrolytic release (PR) or carbon assimilation experiment tests the surface material of Mars for the presence of microorganisms by measuring the incorporation of radioactive CO_2 and CO into the organic fraction of a soil sample. The reasons for believing that martian life, if it exists, would be based on carbon chemistry have been summarized (Horowitz, 1976). The experiment is carried out under actual martian conditions, insofar as these can be attained within the Viking spacecraft, the premise being that, if there is life on Mars, it is adapted to martian conditions and is probably maladapted to extreme departures from those conditions.

The experiment operates as follows: A sample of Mars, consisting of martian atmosphere at ambient pressure and 0.25 cm^3 of soil is placed within the 4-cm^3 test cell of the instrument. Martian sunlight is simulated by a 6-W high-pressure xenon lamp, filtered to remove wavelengths shorter than 320 nm. The radiant energy reaching the test chamber, integrated between 335 and 1000 nm, is approximately 20% of the maximum solar flux at Mars in this spectral interval, or about 8 mw cm^2. The short end of the spectrum is removed to prevent the surface-photocatalyzed synthesis of organic compounds from CO that is induced by wavelengths below 300 nm (Hubbard et al., 1971, 1973, 1975). Except under the special conditions of the photochemical synthesis, these wavelengths are generally destructive to organic matter. It is therefore reasonably certain that, if there are organisms on Mars, they have devised radiation protective mechanisms. Laboratory tests have shown that the experiment detects both light and dark fixation of $^{14}CO_2$ and ^{14}CO by soil microbes (Hubbard et al., 1970; Hubbard, 1977) and the instrument can be operated in either the light or dark mode on Mars. The experiments so far conducted were performed in the light. The option exists to inject water vapor into the incubation chamber, but it was not exercised in these experiments.

At the start of an experiment, 20 μl of a mixture of $^{14}CO_2$ and ^{14}CO (92:8 by volume, total radioactivity 22 μC) is injected into the test cell from a reservoir. The resulting pressure increase is 2.2 mbar over ambient, which, at the Viking 1 landing site, is 7.6 mbar. The martian atmosphere is about 95% CO_2 and about 0.1% CO. The addition of the radioactive gases increases the partial pressure of CO_2 by 28% and that of CO 23-fold.

The test chamber and its contents are illuminated for 120 hr at a temperature that depends on both the ambient martian temperature and the quantity of heat generated within the spacecraft. In the two experiments described, the incubation temp-

eratures were 17 ± 1°C and 15 ± 1°C, respectively, with a brief
upward excursion in the second (control) experiment to 20°C.
This temperature range is clearly above the soil surface temperature at the Viking 1 site, where a maximum of -5°C has been
estimated during these observations (H. Kieffer, personal communication).

At the end of the incubation period, the unreacted $^{14}CO_2$
and ^{14}CO are vented at 120°C from the test chamber and the soil
is heated to 625°C to pyrolyze any organic matter it contains.
The volatile products (including unreacted $^{14}CO_2$ and ^{14}CO desorbed from the walls and soil particles) are swept from the
chamber by a stream of He and introduced into a column of Chro-
mosorb P coated with CuO, which functions as an organic vapor
trap, operating at 120°C. Organic fragments (larger than
methane) are retained by the column, but $^{14}CO_2$ and ^{14}CO pass
through and their radioactivity is counted: this count is referred to as peak 1. The column temperature is then brought to
650°C, releasing organic compounds and simultaneously oxidizing
them to CO_2 by means of the CuO contained in the column packing.
The radioactivity of this $^{14}CO_2$ is called peak 2: it measures
organic matter synthesized from $^{14}CO_2$ or ^{14}CO during the incubation period.

The results are shown in Table III. Experiment 1 was an
active experiment, conducted as described above. Experiment 2
was a control in which a second portion of the same surface
sample was heated to 175°C for 3 hr before the start of incubation. The high background radioactivity comes primarily from
two radioisotopic thermoelectric generators that supply power
to the lander. Counting times were sufficiently long to detect
approximately 10 count/min above this background. The counts
were remarkably free of noise, except during the latter part
of the second experiment when some noisy segments appeared. The
noise was not random since the errors were all in the same (upward) direction. These segments were edited out before the
data were averaged. All the counting rates summarized in Table
III are Poisson distributed.

The "expected" counting rates (Table III) are those predicted if no ^{14}C is fixed into organic matter. These counts
represent the fraction of peak 1 retained at 120°C and eluted
at 650°C. This fraction is known from laboratory tests: when
peak 1 equals 10^4 count/min, the maximum fraction retained is
2×10^3, or 15 count/min for the experiments reported.

Analysis of the results shows that a small but significant
formation of organic matter occurred in experiment 1. The inhibition of this process in experiment 2 shows it to be heat
labile. Until a dark control is completed, we cannot know
whether the fixation is light dependent. The amount of organic
carbon represented by 96 - 15 = 81 count/min is equivalent to
the reduction of 7 pmole of CO or 26 pmole of CO_2. Laboratory

TABLE III. *Pyrolytic Release Counting Rates and Their Standard Errors*

	Counts per minute			
Experiment	Total	Background	Net	Expected
		Peak 1		
1 (active)	7899 ± 59	478 ± 0.62	7421 ± 59	
2 (control)	8129± 60	480 ± 0.57	7649 ± 60	
		Peak 2		
1 (active)	573 ± 0.83	477 ± 0.79	96 ± 1.15	⩽15
2 (control)	500 ± 0.47	485 ± 1.20	15 ± 1.29	⩽15

experience based on terrestrial soils suggests that two or three times more organic matter may remain in the pyrolyzed soil as a nonvolatile tar (Hubbard *et al.*, 1970; Hubbard, 1977).

Although these preliminary findings could be attributed to biological activity, several experiments remain to be done before such an interpretation can be considered likely. In particular, the effect observed in experiment 1 must be confirmed in a second test, and the presence of organic matter in the martian surface must be demonstrated. Given the unusual conditions that prevail at the surface of Mars, the possibility of nonbiological reduction of CO or CO_2 cannot be excluded at this time.

THE GAS EXCHANGE EXPERIMENT

The gas exchange experiment (GEX) measures compositional changes in the atmosphere above a soil sample upon addition of aqueous nutrient medium, and from these data it attempts to show the presence of microbial activity. The results from the first 20 sols of incubation show significant changes in the composition of the experimental atmosphere.

GEX activities that occurred after landing, up to the end of the first incubation cycle, are given in Table I. Descriptions of the concept governing the design of the experiment and results obtained have been described (Oyama *et al.*, 1970, 1971, 1977; Merek and Oyama, 1970).

The first incubation cycle begins with the addition of 1 cm^3 [2] of packed martian soil to the incubation chamber. In the

[2] *Solid volume of soil delivered was estimated to be 0.465 cm^3.*

process of loading the soil and sealing the test cell on sol 8, martian atmosphere was trapped within the chamber at the prevailing pressure. The mixture of Kr, CO_2, and He gases [3] and 0.57 cm^3 [4] of aqueous nutrient medium containing neon were added to the test cell. This amount of nutrient was added to the bottom of the test cell so that the soil sample was contacted by water vapor only, and not by the liquid medium. Results of the analyses of the headspace gases during the humid (water vapor) mode are shown in Table IV. All results are corrected for the initial contributions of the original trapped martian atmosphere: the added Kr, CO_2, and He gas mixture; the trace amounts of gases introduced by the nutrient injection, and losses from sampling the headspace gas. Calculation of the actual gas concentrations is based on their partitioning between the gas and liquid phases at the incubation temperature (Clever and Battino, 1975; Morrison and Johnstone, 1954; Yet and Peterson, 1964; Cos and Head, 1962; Austin *et al.*, 1963; Morrison and Billett, 1962). (Table IV).

The chromatogram shows that carbon dioxide, oxygen, nitrogen, and argon and carbon monoxide (measured as a single peak) are evolved from the soil sample when warmed to 8 to 10°C and humidified. The maximum amount of nitrogen gas, 15 nmole, appears on sol 11 and decreases to one-half of this value by sol 15. Oxygen, on the other hand, after reaching its maximum on sol 11, appears to plateau. If one assumes oxidation of ascorbic acid in the medium, the actual total amount of oxygen produced equals 725 nmole (640 released into the atmosphere plus the 85 nmole consumed in the oxidation of the added ascorbic acid). The maximum amount of CO_2 produced on sol 10 is approximately 9100 nmole, which decreases on sol 11 - 8800 nmole. As is indicated later, the readsorption of CO_2, even after corrections for solubility, is likely associated with basicity changes in the mixture of soil and aqueous nutrient. No conclusion on the presence of CO can be drawn because of the low values of the Ar and CO peak. The values of Ne and Kr demonstrate the consistency of the internal standards and the apparent precision for the gas analyzers.

The anomalous amount of O_2 accompanying the desorption of CO_2 represents an enrichment of 18 times in the martian soil. The results suggest either that molecular oxygen is held in

---

[3] *The composition of the mixture was 5.51% Kr, 2.84% CO_2, 91.47% He, ϕ 0.14% N_2, 0.035% O_2. The GEX test cell temperatures ranged from 8.3 to 10.8°C.*

[4] *Nutrient volume injected into the test cell estimated from the quantity of Ne in the head space above the incubating soil. Neon was added to the nutrient ampule before it was sealed.*

TABLE IV. Gas Composition (Corrected) in Gas Exchange Test
Cell (Humid Mode)

Gas	Gas emitted (nanomoles) after humidification (hours) on Mars date:				
	(2.78) Sol 9	(27.86) Sol 10)	(52.51) Sol 11	(101.91) SOL 13	(150.74) Sol 15
N_2	7	11	16	12	8
O_2	460	610	640	630	630
CO_2	5500	9100	8800	8900	8400
Ar^a	3	2	7	3	1
Ne^b	20	20	18	20	21
Kr^c	2000	2000	2000	2000	1900

[a]The gas chromatograph detector data are sampled at 1-second intervals, digitized, and fitted to a skewed gaussian distribution from which peak heights were obtained. The gas in the headspace is obtained from the ratio of the sample loop volume to the total headspace volume. The cumulative gas composition is corrected for sampling losses by referencing absolute changes in the krypton values for successive samples. Corrections are made for pressure sensitivity in this flight instrument caused by a partial restriction in the gas sampling system which prevents total evacuation of the sample loop to ambient pressure prior to filling (three times) from the test cell. The value for krypton is corrected for pressure as follows:

$$\text{Nanomoles } Kr = 37.77 \ (P_e)^{-0.118} \cdot (V_p)^{1.016}$$

where P_c is the test cell pressure in millibars and V_p is the peak height in volts. The value for each gas is corrected by the ratio of the term $37.77 \ (P_c)^{-0.118}$ to the similar Kr value from a pressure insensitive instrument. The gas composition as stated is corrected by removal of contributions from known sources (for example, trace contaminants in injected gases) and for the values and temperature coefficients. The effects of pH on the CO_2 distribution are included by estimating changes in apparent CO_2 levels on nutrient injections in LR (second injection) and on the wet-mode nutrient injection in gas exchange. The relationship used is

$$\frac{(nanomoles, \ dissolved)}{(nanomoles, \ gas \ phase)} = L \ \frac{(volume, \ liquid}{(volume, \ gas)}$$

where the L values for CO_2 are sols 9 and 10.21.4; on sols 11 to 15.28.4; on sol 16, 40.4; and on sols 17, 18, 20, 25, and 28, 68.5.

[b]*Assumed to be Ar as Ar is not resolved from CO on this column.*

[c]*Mean value for Ne, 19.88 ± 0.95 (4.80%); mean value for Kr, 1976 ± 21.54 (1.09%).*

relatively large quantities in the martian soil and released upon warming in the incubation test cell or that oxygen is generated from some unstable oxidant upon warming or, more likely, upon contact with water vapor.

During the entire first cycle, no H_2, NO, or CH_4 was detected in the headspace. The absence of hydrogen upon wetting the soil seems to preclude the presence of metallic iron in concentrations greater than 0.003%.

Absorption of CO_2 at martian surface temperatures and desorption at the incubation temperature of the test cell could account for some of the desorption during the 21.23 hr that the soil was sealed in the test cell. However, the data suggest that the major desorption of the CO_2 occurred in the 2.78 hr immediately after the humidification of the test cell. These points remain to be investigated in the laboratory under similar conditions.

On sol 16, an additional 2.27 cm^3 of nutrient was injected. Including the amount added earlier, the nutrient now measures 2.84 cm^3, and wets the soil. The data for the wet mode are shown in Table V.

The decrease in CO_2 seen immediately after wetting the soil may be due to pH changes of the soil-aqueous solution mixture. The slow rise in CO_2 content of the atmosphere after this initial decrease is not readily explained. This could be the result of further changes in this pH of the wet soil, or the oxidation of some of the substrates in the medium by the oxidants postulated above. That the CO_2 arises as a result of biological oxidation cannot, of course, be ruled out at this time. The decrease in oxygen can be accounted for by the additional ascorbic acid in the fresh nutrient added on sol 16.

The changes observed in the N_2 content of the incubation atmosphere are minimal and may be explained by a number of processes including sorption by the soil, or by Van Slyke reactions between the α-amino acids of the medium with residual nitrites in the soil. On the other hand, a biological origin (denitrification of added nitrates in the medium) is also possible.

TABLE V. Gas Composition (Corrected) in Gas Exchange Test Cell (Wet Mode)

Gas	Gas emitted (nanomoles) after 2.3 cm^3 of nutrient injection (hours) on Mars date					
	(2.66) Sol 16	(27.31) Sol 17	(51.98) Sol 18	(100.31) Sol 20	(223.88) Sol 25	(295.21) Sol 28
N_2	-6	-5	-5	-4	2	4
O_2	460	380	270	210	20	210
CO_2	8400	9500	10400	10800	10800	10000
Ar^a	-3	-3	-4	-3	-3	-3
Ne^b	99	150	160	160	160	170
Kr^c	1400	1400	1400	1400	1400	1400

[a] As in Table IV.
[b] As in Table IV.
[c] As in Table IV.

THE LABELED RELEASE EXPERIMENT

The labeled release (LR) experiment (Horowitz et al., 1972; Levin and Straat, 1977) seeks to detect metabolism or growth through radiorespirometry (Levin, 1963). The radioactive nutrient used for the test consists of seven simple organic substrates (formate, glycolate, glycine, D- and L- alanine, D- and L-lactate), each present at $2.5 \times 10^{-4}M$ and each equally and uniformly labeled with ^{14}C (8μC/μmole).

To initiate the LR experiment on Mars, 0.5 cm^3 of the sample was placed inside a test cell, which is connected by a tube (33 by 0.2 cm, inside diameter) to another chamber flanked with two solidstate beta detectors. The background radioactivity, caused primarily by the radioisotopic thermoelectric generators powering the lander, was counted for approximately 24 hr prior to nutrient injection and found to be 490 count/min. The sample was then injected with 0.115 ml of the radioactive nutrient. This volume of nutrient contains approximately 257,000 count/min, each of the 17 carbons of the seven substrates contributing

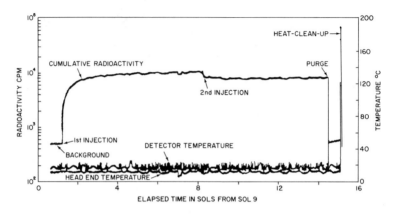

FIGURE 1. Plot of labeled release data from first analysis on Mars. Radioactivity was measured at 16-minute intervals throughout the analysis cycle, except for the first 2 hr after the first nutrient injection when readings were taken every 4 minutes. Detector and head-end temperatures were measured every 16 minutes.

approximately 15,000 count/min (corresponding to 29 nmole of carbon). Approximately 7 sols after the first nutrient injection (Table I), a second nutrient injection was made, and incubation was continued for an additional 6 sols. After each nutrient addition, radioactive gas evolved into the headspace above the sample equilibrated with the gas volume in the detector chamber. The gas accumulating within the detector chamber was continuously monitored for radioactivity during the incubation period. The temperature of the detector and the head end of the test cell were also monitored throughout the cycle. At the end of this incubation, a cycle was conducted with a second 0.5 cm³ portion of the original sample held in reserve in the lander for this purpose. This was placed in a clean test cell, sealed, and heated at $170^{0}C$ for 3 hr. After the cell cooled and background had been counted for approximately 20 hr, nutrient was injected, and the evolved radioactive gas was compared to that from the first analysis. Details of the nutrient, instrumentation, and terrestrial assays have been described (Levin and Straat, 1977).

Upon injection of the labeled nutrient on sol 10, a vigorous production of radioactive gas was observed in the test cell as shown in Fig. 1, where data for the entire first cycle of the experiment are presented. The initial course of evolution of gas resembled that displayed by microbiologically active terrestrial soils (Levin and Straat, 1977). However, the rate of evolution of radioactive gas from the martian sample slowed more rapidly than would have been expected for a terrestrial soil, and approached a plateau of approximately 10,000

count/min over background. The magnitude of the response
corresponds to approximately 65% of one of the labeled carbons
in the nutrient. These facts could be an indication that only
one of the substrates may have been involved in the reaction.

Upon addition of a second volume of labeled nutrient on sol
17, an immediate (within 10 minutes) increase in evolution of
radioactive gas was followed by a rapid decrease of radioactivity until a new plateau was reached at approximately 8000
count/min. This decline accounts for approximately one-third
of the total amount of gas that had been evolved, including the
spike (Fig 2), which appears immediately after the commanded
nutrient injection. However, after reaching plateau, the
radioactivity level slowly rose over the ensuing 6 sols at an
average rate of approximately 40 count/min per sol. This rate
is considerably less than that observed following the first injection.

In isolating the biology instrument against the martian
diurnal temperature fluctuation (approximately 187 - 242°K) at
the landing site, the thermal environment shown in Fig. 1 was
imposed upon the LR module by the instrument temperature control system. Thus, the head end fluctuated between 9 and 13°C,
and the detector temperature cycled between 14 and 26°C. Minor,
regular patterns of fluctuation in the radioactivity curve correlate with the temperature of the test cell. Such fluctuations
were anticipated and are not indicative of instrument anomalies.

Thirteen sols after the first injection, cycle 1 of the LR
experiment was terminated. To remove the accumulated radioac-

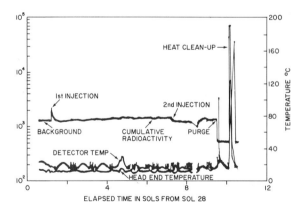

*FIGURE 2. Plot of labeled release data from control anal-
ysis on Mars. Radioactivity was measured at 16-minute intervals
throughout the cycle, except for the first 2 hr after each nutrient injection when readings were taken every 4 minutes. Detector and head-end temperatures were measured every 16 minutes.*

tive gas and dry the test cell, the detector and test cell were
purged with helium. A clean test cell was then rotated under
the head end, and both detectors and head end were heated during continuous helium purging to minimize the remaining radioactivity. Background was then counted for about 20 hr. The
new background level after the analysis averaged 516 count/min
compared to the average of 490 count/min prior to the first injection.

Because of the positive response in cycle 1, a control sequence was run in cycle 2. After the control sample was heated
(as described earlier), the test cell was vented to equilibrate
its headspace with the martian atmosphere. After venting, the
radioactivity was observed to be 1300 count/min (including the
516 count/min background), a baseline level not expected to in-
terfere seriously with the experiment.

After acquisition of the surface sample, nutrient was delivered to the heat-treated sample. The ensuing control data
are shown in Fig. 2. Some immediate release of radioactive
gas, totaling approximately 800 count/min above the new baseline of 1300 count/min, occurred. However, the released gas
immediately began to disappear from the detector cell, and,
within about 8 hr, the radioactivity was virtually at the baseline level of 1300 count/min. After this, a slight rise in
radioactivity was observed, less than that seen in the latter
part of the commanded injection phase of cycle 1.

Because most terrestrial control soils sterilized by heat
demonstrate an immediate, low-level release of radioactive gas
that quickly reaches a plateau and remains constant, the possibility was considered that the decline in radioactivity seen in
Fig. 2 resulted from a gas leak in the test cell. The data obtained during background counts prior to the control show that
the 1300-count/min baseline purged down to the approximate initial 516-count/min background level. Thus, radioactive gas was
responsible for the elevated baseline prior to the first injection. If there was a leak, a reduction in the 1300 count/min
would have been observed before the injection.

DISCUSSION

The experiments described above give clear evidence of
chemical reactions. The essential question is whether they are
attributable to a biological system. We are unable at this
time to give a clear answer to that question, partly because
the planned experimental program is not yet completed, and
partly because of the inherent difficulty in defining complex
living organisms which may have developed and evolved in an environment completely different from that of the planet Earth.

An important consideration in evaluating the possibility of life on Mars is the chemical analysis of carbon compounds in the martian soil. Biemann *et al.* (1976) reported that no organic compounds larger than methanol and propane, for example, were observed in the Viking I samples at detection limits that range from 0.1 to 50 parts per billion. The results are somewhat similar to those found in an Antarctic soil (No. 542, collected by R. E. Cameron) that has little organic material and appears not to support an active biota (Horowitz *et al.*, 1972). These results, especially if reinforced by analyses at a second martian site, would tend to make biology on Mars less likely, at least in the terrestrial mode.

It is difficult to compare directly the results of the three biology experiments since each was conducted under different conditions. Nonetheless, it is interesting that the two experiments dealing directly with radioactive carbon chemistry yielded positive responses, and both were eliminated by heat sterilization of the martian sample.

These results violate none of the prima facie criteria for a biological process, and show some of the most general characteristics of known organisms. The positive result of the PR experiment signifies the reduction of CO or CO_2, or metabolic exchange with reduced organic compounds, which are exhibited by all terrestrial organisms. On the other hand, nonbiological photoreduction of CO can also be demonstrated at shorter ultraviolet wavelengths (Hubbard *et al.*, 1971, 1973, 1975), and catalytic dismutation of CO is also well established.

In contrast, the LR experiment requires conversion of oxidizable substrates into radioactive gas. In a terrestrial test, the collective results of a positive response in cycle 1 and its elimination by heat sterilization in cycle 2 would support the concept that microorganisms were present in the sample. The amplitude of the test response is an order of magnitude above that expected from a sterile soil, and the difference between the Mars test and the control cycle exceeds the 3σ level, which has been chosen as a criterion for a positive response (Levin and Straat, 1977). However, important caveats to such a conclusion are (1) the possible limitation of metabolism to one substrate and (2) the lack of an exponential phase of gas evolution indicative of growth. Organisms in terrestrial soils attack more than one substrate, as evidenced by the fact that the plateaus attained generally represent 50% or more of the total label added (Levin and Straat, 1977). On Mars, however, utilization of only one of the offered terrestrial substrates might indicate a selective metabolism. The abrupt change in environmental conditions of the martian soil imposed by the biology instrument with respect to water and temperature, together with the relatively short time of the experiment, might readily account for lack of growth. The absence of a positive response to the second injection in cycle 1 similar to that seen from the first injection might be attributed to inhibition or death of the microorganisms.

Despite the suggestive character of these responses of the Mars sample, the environmental conditions on Mars are sufficiently different from those on Earth to require cautious interpretation. A high ultraviolet flux strikes the martian surface material, and may result in the production of highly reactive compounds capable of oxidizing the labeled nutrient. However, any explanation must account for the kinetics of the reaction as well as the heat lability of such oxidants or catalysts at 170 to 175°C. Similarly, the absorption of radioactive gas after the second injection of nutrient may be facilitated by alkalinity induced in the martian soil by wetting. An absorption of CO_2 was also seen in the GEX upon wetting the sample.

Final interpretation of the results must await the results from the investigations on the second lander, the completion of Viking 1 studies, and ground-based laboratory experiments.

ACKNOWLEDGMENTS

We acknowledge the effective and tireless efforts of the engineers who are part of the Viking Biology Flight team and without whom these experiments could not have been accomplished. R. I. Gilje is chief engineer on this team, which also includes S. Loer, C. Reichwein, G. Bowman, and D. Buckendahl. Supported, in part, by NASA contracts NASI-9690 (to G.V.L.), NASI-12311 (to N.H.H.), NASI-13422 (to J.S.H.), and by NASA grants NGR-050 02308 (to N.H.H.) and NSG-7069 (to J.S.H.). We also acknowledge the profound contribution to the Viking mission and particularly to the development of life detection concepts, of our former colleague and deputy team leader, Dr. Wolf Vishniac. His untimely accidental death in Antarctica in 1973 deprived us of his keen insight and inquiring mind at a crucial time in this study. The invaluable assistance of Dr. Richard S. Young in the preparation of this manuscript is also acknowledged.

REFERENCES

Austin, W. H., Lacombe, E., Rand, P. W., and Chatterjee, M.
(1963). *J. Appl. Physiol. 18,* 301.
Biemann, K., Oro, J., Toulmin, P. III, Orgel, L. E., Nier, A.
O., Anderson, D. M., Simmonds, P. G., Flory, D., Diaz, A.
V., Rushneak, D. R., and Biller, J. A. (1976). *Science*
194, 72.
Clever, H. L., and Battino, R. (1975). *In* "Solutions and Solu-
bilities" M. R. J. Dack, ed.), Partl, Chap. 7. Wiley, New
York.
Cos, J. D., and Head A. J. (1962). *Trans. Faraday Soc. 58,*
1839.
Hall, L. B. (1975). *In* "Foundations of Space Biology and Medi-
cine" (M. Calvin and O. G. Gazenko, eds), Vol. 1, pp. 403-430.
NASA, Washington, D. C.
Horowitz, N. H. (1976). *Accounts Chem. Res. 9,* 1.
Horowitz, N. H., Hubbard, J. S., and Hobby, G. L. (1972). *Ic-*
arus 16, 147.
Horowitz, N. H., Cameron, R. E., and Hubbard, J. S. (1976).
Science 176, 242.
Hubbard, J. S. (1977). *Origins Life* in press.
Hubbard, J. S., Hobby, G. L., Horowitz, N. H., Geiger, P. J.,
and Morelli, F. A. (1970). *Appl. Microbiol. 19,* 32.
Hubbard, J. S., Hardy, J. P., and Horowitz, N. H. (1971). *Proc.*
Natl. Acad. Sci. U.S.A. 68, 474.
Hubbard, J. S., Hardy, J. P., Voecks, G. E., and Golub, E. E.
(1973). *J. Mol. Evol. 2,* 149.
Hubbard, J. S., Voecks, G. E., Hobby, G. L., Ferris, J. P.,
Williams, E. A., and Nicodem, D. E. (1975). *J. Mol. Evol.*
5, 223.
Klein, H. P. (1974). *Origins Life 5,* 431.
Klein, H. P., Lederberg, J., Rich, A., Horowitz, N. H., Oyama,
V. I., and Levin, G. V. (1976). *Nature (London) 262,* 24.
Levin, G. V. (1963). *Adv. Appl. Microbiol. 5,* 95.
Levin, G. B. (1972). *Icarus 16,* 153.
Levin, G. V., and Straat, P. A. (1977). *Origins Life,* in press.
McDade, J. J. (1971). *In* "Planetary Quarantine: Principles,
Methods, and Problems" (L. B. Hall, ed.), pp. 37-62. Gordon
& Breach, New York.
Merek, E. L., and Oyama, V. I. (1970). *Life Sci. Space Res. 8,*
108.
Morrison, T. J., and Johnstone, N. B. (1954). *J. Chem. Soc.,*
3441.
Oyama, V. I. (1972). *Icarus 16,* 167.

Oyama, V. I., Merek, E. L., Silverman, M. P., and Boylen, C. W.
   (1971).
Oyama, V. I., Berdahl, B. J., Boylen, C. W., and Merek, E. L.
   (1972). *Proc. Lunar Sci. Conf. 3rd* (C. Watkins, ed.), p.
   590.   Lunar Science Institute, Houston, Texas.
Oyama, V. I., Berdahl, B. J., Carle, G. C., Lehwalt, M. E.,
   and Ginoza, H. S. (1977). *Origins Life* in press.
Yet, S. Y., and Peterson, R. E. (1964). *J. Pharm. Sci. 53*, 822.

3

THE CURRENT STATUS OF SPECULATIONS ON THE COMPOSITION OF THE CORE OF THE EARTH[1]

Robin Brett

U.S. Geological Survey, Reston, Virginia

The earth's core, at a pressure of between 1.3 and 3.7 Mbar and a temperature between 3000° and 5000°C, consists primarily of Fe-Ni in the inner core and molten Fe alloyed with from 8 to 20% of a light element, most likely S or Si, giving an average atomic number for the elements in the outer core of about 23. If the core accreted before the mantle, then the apparent disequilibrium of some elements with respect to core and mantle is explained, but it is difficult to explain the presence of appreciable quantities of a light element that are required in the core. Alternatively, the core·formed by melting, coalescence, and subsequent collapse of metallic particles in a protoearth that was relatively homogeneous initially. Such collapse would raise the temperature of the earth as much as 2300° K. Silicon must be present in the core if (Fe + Mg)/Si > 1 in the mantle and if that ratio for the bulk earth lies within the limits set by the sun and chondrites. If Si is present, the apparent chemical disequilibrium between metallic Si in the core and Fe^{2+} and Fe^{3+} in the mantle must be explained. Arguments for S in the core based on the abundance of S in rocks of crustal and mantle origin appear to have little foundation; however, there are strong geochemical and cosmochemical arguments justifying the presence of S in the core if the earth accreted at relatively low temperatures. The intriguing speculation that the bulk of the earth's potassium is present in the outer core, thus providing energy for core and mantle, remains to be tested by more partitioning experiments. A number of explanations

---

[1]*Reprinted with permission from Reviews of Geophysics and Space Physics 14, 375-383, copyrighted by American Geophysical Union.*

*exist to explain the apparent disequilibrium of some elements
between core and mantle; however, none is satisfactory. The
core formed soon after the earth accreted; in fact, the age
generally taken as the age of the earth probably represents
the time of core formation.*

INTRODUCTION

In the latter part of the 19th century, shortly after it
was recognized that meteorites can be divided into two broad
classes, stony meteorites and iron meteorites, there were suggestions that the earth may have an iron core surrounded by a
silicate shell (e.g., Dana, 1875). Weichert (1897) refined this
hypothesis by pointing out that it explained the high mean density of the earth relative to silicates. The presence of a
core was confirmed on the basis of seismic properties by Oldham (1906). Seismologists soon accepted the concept that the
earth has a core and that it is made predominantly of iron.

If the ratios of nonvolatile elements in the earth are similar to those in the sun and chondritic meteorites, then an
iron-rich core is required through considerations of elemental
abundances. A number of workers, most notably Ramsey (1948)
and more recently Lyttleton (e.g., 1973), have challenged the
concept of an iron core, suggesting that the mantle silicates
undergo phase changes in the earth's deep interior to produce
material of high density, low melting point, and high electrical conductivity. Ramsey's hypothesis is accepted by few geo-
physicists today for a number of reasons, not the least being
that phase changes in a multicomponent system would be expected
to occur over a range of depths, yet the core-mantle boundary
has been shown to be a very sharp discontinuity. Recently,
Bird and Weathers (1975) state that josephinite, a nickel-iron
alloy intergrown with a magnesium silicate phase, andradite,
arsenide and sulfide phases and alleged native Si and 'CaO,2FeO'
may have been transported from the core by a deep mantle plume.
The material occurs as placers in a stream in Oregon. Until
the mineralogy of the material can be thoroughly documented, its
geological setting determined, and evidence for a very high
pressure origin established, the claim must be dismissed as un-
fettered speculation.

Recently, there has been considerable interest in the nature
of the light element(s) present in the core. Detailed knowledge
of core chemistry should provide important new information on
the bulk composition of the earth and on the thermal regime

within the core and provide new insight into the nature of the core-forming and accretional processes. The present paper attempts to review our knowledge of core chemistry, since it has been 5 yr since the excellent discussion on core chemistry and physics by Anderson *et al.* (1971).

TEMPERATURES AND PRESSURES WITHIN THE CORE

Pressure estimates for the interior of the earth are virtually independent of compositional models and are known reasonably accurately, since they rely largely on estimates of mass, coefficient of moment of inertia, and seismic velocity data. Temperatures in the interior are less understood than pressures and are probably not known to better than about ±30% because estimates rely on an understanding of the adiabatic gradient and *P-T* melting curves, which are dependent on composition.

A major pioneering step in the calculation of densities throughout the earth was made by Bullen (1936). He fitted mass and moment of inertia of a model to those values observed for the earth. Clark and Ringwood (1964) calculate pressures of 1.39 Mbar and 1.36 Mbar for the core-mantle boundary for pyrolitic and eclogitic earth models, respectively. They calculate pressures at the center of the earth to be 3.57 and 3.68 Mbar, respectively.

Temperatures can be very approximately estimated for the core by the extrapolation of melting curves to core pressures (e.g., Sterrett *et al.*, 1965; Kraut and Kennedy, 1966; Higgins and Kennedy, 1971; Usselman, 1975). If we can assume that the inner core-outer boundary represents a phase change from solid to liquid iron, then we can estimate the temperature at that point if we know the melting curve of iron. The outer core must be above the melting point, so the melting curve for iron as a function of pressure would represent a minimum gradient for the outer core if the core were pure iron. However, the outer core is not pure iron, and the effect of even small amounts of some additional components (e.g., C.S) on the melting point of iron is considerable (e.g., Hansen and Anderko, 1958). A number of methods exist for extrapolating data concerning the effects of pressure on melting temperature (e.g., Kraut and Kennedy, 1966), but considering the immense pressure in the core, and hence the extrapolations required, considerable errors must occur in any extrapolation. Also the solid inner core must consist of a high-pressure modification of iron. The melting temperature of a high-pressure polymorph may be several hundred degrees above the melting points of the lower-pressure forms of iron whose melting curves have been determined in the

laboratory, thus causing further uncertainty in temperature
estimation (Birch, 1972).

Rather than discussing the various temperature estimates
that have been made and the methods and assumptions used to
determine them, for the sake of brevity it is best to agree
with Birch (1972), who states that based on the melting curve
of pure iron all we can say is that the core-mantle boundary
must be at a temperature of less than 4000°C and the center of
the earth must be less than 5000°C.

MECHANISM OF CORE FORMATION

The nature of the light element(s) present in the core
should be closely related to the core-forming and accretional
processes. By accepting a given accretional model, the nature
of the core-forming process and core chemistry are immediately
constrained. A logical and important consequence is that resolution of the light element problem will invert to an important
constraint on early earth history. So far as core formation is
concerned, accretional models may be subdivided into two groups:
(1) those in which the earth's metallic core condensed and
accreted more or less homogeneously and the core formed by coalescence and sinking of metallic particles within the mantle.
Ringwood (1975) has a thorough review of the various accretional models and models of core formation. Models of core formation and composition are listed in Table I.

Primary Condensation Models

Eucken (1944), Turekian and Clark (1969), and Clark *et al.*
(1972) propose that due to the higher condensation temperature
of Fe with respect to magnesium silicates in the solar nebula,
a metallic Fe-Ni core accreted first and was followed by silicates which formed the mantle. The proposal does not require
an explanation of why the core and mantle appear to be in chemical disequilibrium (Ringwood, 1966a,b), since disequilibrium
is implicit in such a model. Also heat is not required early
in earth history to melt the protocore to allow coalescence of
core material within the mantle. The model has the following
problems associated with it, however:

1. At a pressure of 10^{-3} atm (a figure near the upper limit
of pressures pertaining during condensation in the nebula)
(Grossman and Clark, 1973), metallic Fe condenses only about
$10°-20°C$ above the condensation temperature of forsterite
(e.g., Grossman and Larimer, 1974). At lower pressures the

condensation temperatures approach one another until finally forsterite condenses at a temperature above that of the condensation temperature of Fe. There are sufficient uncertainties in the thermodynamic data used for condensation calculations that forsterite may condense at a higher temperature than Fe over the pressure range of interest, thus destroying the model. For example, SiO is an important component in the calculations, since it is the main silicon species assumed present in the nebula (e.g., Grossman, 1972). The more reliable estimates of the heat of formation of SiO spread over a range of about 3 kcal/mol (Stull *et al.*, 1971).

2. The core was molten early in earth history (see below). The condensation temperature of metallic Fe at the above pressure is well below the melting point of Fe (1534°C at 1 atm (Hansen and Anderko, 1958)). The earth must have therefore accreted relatively rapidly for the core to have become molten early; slow accretion would result in extensive radiative cooling. Anderson and Hanks (1972) evade the thermal problem by suggesting that refractory compounds (Ca, Al, Ti oxides and silicates) that calculations show would condense from the solar nebula at higher temperatures than Fe (e.g., Grossman and Larimer, 1974) accreted first. Metallic Fe-Ni then accreted around these phases, followed by the phases that now constitute the mantle. Anderson and Hanks state that the refractory compounds would be enriched in the refractory elements Th and U, thus providing a deep-seated heat source. As the metallic Fe-Ni became molten, it displaced the refractory oxides which would then homogenize with the mantle.

3. Condensation theory requires that the metal condensing be almost pure Fe-Ni-Co (e.g., Grossman and Larimer, 1974). Density considerations require that there be considerable amounts of a light element in the core (see below). No provision exists in the model for the concentration of such an element in the core.

Postaccretional Differentiation Models

The more conventional models of core formation postulate that the earth accreted homogeneously from nebula condensates, including particles of Fe-Ni, or metal + FeS (e.g., Lewis, 1973) which were disseminated throughout (e.g., Urey, 1952). Core formation occurred by sinking and coalescence of molten metal. The outer portion of the earth must have been hot, possibly well above silicate solidus temperatures soon after accretion, since more and more data indicate that the outer portion of the moon, which is a much smaller body than the earth, was probably above solidus temperatures at the time of formation (e.g., Toksöz *et al.*, 1973; Taylor and Jakes, 1974). Elsasser (1963) in

TABLE I. Summary of Core Formation Hypotheses

Model	Principal authors
Primary accretion or condensation hypothesis	Eucken (1944), Turekian and Clark (1969), Clark et al. (1972)
Allende accretion hypothesis	Anderson and Hanks (1972)
Cool accretion	Urey (1952), Elasser (1963)
Hot accretion with early core separation	Hanks and Anderson (1969), Ringwood (1960), Oversby and Ringwood (1971)
Hot accretion with early core separation	Ringwood (1966a,b)
Hot accretion with S in protocore	Murthy and Hall (1970)

Mechanisms	Strengths	Weaknesses
Heterogeneous Accretion Fe-Ni accretes directly from nebula due to higher-condensation T; silicates follow.	There is no core-mantle disequilibrium problem and no problem in core formation.	Early deep heat source is required to cause melting; light element cannot accrete with Fe-Ni; thermodynamic data for condensation theory need refining.
Refractory Ca,Al,Ti oxides and silicates (U-, Th-rich) accrete first, followed by Fe, Ni, and then silicate mantle; interior heats; protocore sinks to center.	As above; early deep heat source is provided.	As above, except heat problem is removed.
Homogeneous Accretion Cool accretion of homogeneous earth takes place; radioactive heating leads to core formation with consequent heat evolution. Early intense accretional heating caused early core formation.	Light element can be alloyed with Fe-Ni before or during core formation; no preaccretion differentiation is required. As above; early core formation is possible.	Is core mechanism possible? There is lack of early heat source; apparent core-mentle disequilibrium must be explained. As above, except heat source problem is removed.
Deep interior accreted cool; outer parts were heated, causing reduction by C to Fe-Si and massive CO blowoff; core then formed.	As above; there is Si in core.	Tremendous CO loss must occur from earth; apparent core-mantle disequilibrium must be explained; does earth contain enough Si?
As above, except low T of Fe-FeS eutectic results in early formation of Fe-FeS core; there is no Si in core, no massive reduction.	S appears to be most reasonable geochemically and cosmochemically; there is low T of core formation.	Why did S accrete in appreciable quantities and not alkalies?

his classic paper on the early thermal history of the earth
suggests that viscosity in the mantle increases considerably
with depth, since the adiabatic temperature-pressure gradient
in the mantle is much smaller than the slope of the curve of
liquidus temperature versus pressure. The denser molten core-
forming material would sink down through the solid mantle,
forming a layer at a depth where the viscosity became too great
for further sinking. Small undulations at the base of this
layer would tend to grow under the influence of gravity in much
the same way that a salt dome grows. Finally, the structure
would become unstable and collapse to form the core. As Birch
(1965) remarked, 'It is evidently a highly unsymmetrical process, somewhat reminiscent of the parlor trick of removing the
vest without the coat, and it cannot be well followed in detail.
Mantle convection may well have assisted core formation. Due
to the sinking of dense material, gravitational energy would
thus be transformed to thermal energy, which would lead to a
mean increase in the temperature of the earth of about $2300^{\circ}K$
(Flaser and Birch, 1973). The addition of appreciable quantities of a light element like S to the protocore in solution
with the denser elements would lower this increase in temperature. The core-forming process is thus greatly exothermic and
thus escapes the thermal problems of the model involving heterogeneous accretion. It is therefore probable that core formation was the most important event in the earth's thermal history,
and the supposed age of the earth may in fact be the time of
core formation (Oversby and Ringwood, 1971).

Core formation must have occurred rapidly (Ringwood, 1960).
Hanks and Anderson (1969) state that the core must have been
formed before the oldest known rock possessing remanent magne-
tism, since the earth's core is responsible for the magnetic
field. The immense thermal consequences of core formation suggest that the core formed before the oldest known surface rock.
Hanks and Anderson calculate that a short accretion time of the
order 5×10^5 yr to provide the heat is necessary to achieve
such rapid core formation. Ringwood (1960) and Oversby and
Ringwood (1971) point out that core formation must have occurred
soon after accretion. The calculated age of the earth (T^0 is
about 4.6 b.y.) is based on the assumption that the lead/uranium
ratio in the crust and upper mantle had been established at T^0.
Oversby and Ringwood find that Pb favors metallic Fe-rich alloys
over silicates at elevated temperatures, thus suggesting that
appreciable lead was scavenged by the ptotocore during core
formation and is presently in the core. They therefore conclude
that the Pb-U clock in the mantle was reset as a consequence of
core formation and that the 4.6-b.y. age reflects that event.
They suggest that the time between accretion and core formation
was relatively short and was between 10^8 and 5×10^8 yr. Iso-

topic studies indicate that the moon and most meteorites were
formed about 4.6 b.y. ago, which is consistent with Oversby and
Ringwood's conclusion if one assumes that the solar system was
formed at that time. Their conclusion that core formation
happened soon after accretion demands that the earth was hot
very early in its history.

The earth therefore probably accreted rapidly in order to
be hot enough for core formation to occur, in agreement with
Hanks and Anderson's conclusion. It appears that theories of
early earth history that prevailed until recently and that regard the earth as having accreted in a cool, unmelted state
with subsequent liberation of radioactive heat to cause much
later core formation appear unfounded. There is certainly
sufficient accretional energy to cause early heating if accretion were rapid. If all gravitational energy could be converted to thermal energy, then temperatures in excess of $10,000°K$
would be obtained (Ringwood, 1966a).

The other alternative is that the earth accreted more slowly, say, over 10^8 yr, and that heat in addition to accretional
energy was provided by a T-Tauri phase of the sun and radioactive heating during the accretional period. Core formation occurred soon after accretion ceased.

Early core formation and the required high temperatures
suggest that the earth became chemically differentiated very
early in its history and the enrichment in certain elements
(e.g., Ba, Sr, Zr, Be) that we see in the outer portion of the
earth (e.g., Ringwood, 1966b) probably occurred at that time.
It is noteworthy that similar processes of early moonwide differentiation are presently being postulated for the moon by
several workers (e.g., Taylor and Jakes, 1974).

COMPOSITION OF THE CORE OF THE EARTH

As we discussed briefly above, it is now generally believed
that the core of the earth is composed predominantly of metallic iron with the inner core solid and the outer core liquid.
On the basis of density considerations and by analogy with iron
meteorites, the concept of a nickel-iron core became generally
accepted by the early part of this century. This belief was
partly based on the hypothesis, now found to be incorrect (e.g.,
Wasson, 1974), that iron meteorites represent portions of the
core of a fragmented planet. Birch (1952), in a classic paper
in which he calculated density distribution throughout the
earth, found that a core of pure Fe is about 10-15% more dense
than the calculated density of the core. He therefore suggested that iron is alloyed with a light element in the core. A

number of other workers, on the basis of calculations and
measurements involving hydrodynamic sound speed and compressi-
bilities at core pressures, have confirmed and refined Birch's
conclusion (e.g., Balchan and Cowan, 1966; Press, 1968; Mc-
Queen and Marsh, 1966; Al'tshuler et al., 1962; Birch, 1963).
Both shock wave experiments and seismic velocity measurements
suggest that the outer core contains about 20 wt % of a light
element (Press, 1968; Al'tshuler, 1971; Stewart, 1973) and
the inner core may have up to 50 wt % Ni alloyed with it. The
presence of Ni is reasonable both geochemically, since Ni alloys
readily with Fe, and cosmochemically, based on solar and meteoritic abundances of Ni. Assuming the cosmic Ni/Si ratio, the
whole core should contain about 4 wt % Ni (Brett, 1971). However, the earth may not be of cosmic composition. The inner
core should contain about 4 wt % Ni (Brett, 1971). However,
the earth may not be of cosmic composition. The inner core is
only 5 wt % of the mass of the core. If the outer core also
contains Ni, then the amount of the light element must be increased accordingly. Buchbinder (1972) postulates that the
difference in density between the outer and inner core lies between 0.6 and 1.2 g/cm^3. MacDonald and Knopoff (1958) suggest
that the mean atomic number of the outer core is about 23 based
on calculations using the Thomas-Fermi-Dirac equation of state.

The difference in composition between the outer and inner
cores raises interesting alternatives regarding the nature of
the transition. Is the phase change between solid and liquid
simply a consequence of the higher pressures deep within the
core, or is it largely a consequence of the fact that the light
element, which is preferentially dissolved in the liquid phase,
lowers the melting point with respect to the Fe-Ni inner core?
Some seismologists, on the basis of finite strain theory applied
to the density and compressional velocity of the outer core,
suggest that the outer core may not be homogeneous but layered
(e.g., Anderson et al., 1971). Such chemical layering is not
only possible but even probable over such a range of pressure
and temperature. Such liquid layering could occur only if convection was of minor significance in the outer core. Lack of
convection would, of course, have important consequences on the
origin of the earth's magnetic field. Clearly, knowledge of
the light element in the outer core would contribute information
not only on the bulk chemistry of the earth but also on the
accretionary process, nature of core formation, and temperatures
within the core today.

NATURE OF THE LIGHT ELEMENT

A number of light elements have been proposed as alloying
with Fe in the core. These are C, Si, H (Birch, 1952); Si
(MacDonald and Knopoff, 1958; Ringwood, 1959, 1961, 1966a,b);
C. S (Urey, 1960); C, S, Si (Clark, 1963); Si, O, S (Birch,
1964); S (Mason, 1966; Murthy and Hall, 1970, 1972; Lewis,
1973); Mg, O (Adler, 1966); and O (Bullen, 1973). Other light
elements have not been considered because they do not alloy
readily with Fe, because they are not sufficiently abundant
cosmochemically, or because they tend to form oxides or silicates rather than enter the metallic phase. Ringwood (1966a)
rejects H, N, and C because they form interstitial solutions
with Fe and this would increase the density of the core. These
three elements may also probably be eliminated as candidates,
as they are too volatile to have been trapped in appreciable
quantities during accretion. For example, in chondrites, abundances of H_2O (the main H-containing species), C, and N decrease markedly with increasing metamorphic grade. They cannot
totally be dismissed from consideration, however.
Magnesium has much greater affinity for oxygen than for
metallic iron, so it is likely to remain in the mantle. Unless
extreme pressure markedly increased the solubility of MgO in
Fe, as suggested by Adler (1966), Mg can be eliminated as a candidate. Bullen (1973) has suggested that liquid Fe_2O forms the
outer core. The existence of such a compound at high pressures
has not been demonstrated experimentally; the hypothesis therefore is extremely speculative but merits further study.
The remaining candidates for the light element in the core
are Si and S; these are currently the only elements being considered by most workers. Since we have little knowledge of
either phase transitions at core depths or of outgassing conditions in the primitive earth, it would be premature to reject
elements other than S and Si out of hand. However, since
current thinking is focused on these two elements and there
appear to be more data justifying them as the light element
than the others, these elements will be discussed in greater
detail.

Silicon

MacDonald and Knopoff (1958) and Ringwood (1958, 1959, 1961,
1966a,b) have presented calculations supporting the concept that
Si is the light element present in major abundance in the core.
Their conclusion applies if (1) the mantle is such that (Fe +
Mg)/Si > 1, where the ratio is in atomic percent (this merely

implies that the mantle is more mafic than tholeiitic basalt
or eclogite), or (2) the $(Mg + Fe)/Si$ ratio for the whole earth
is less than 1.65, the ratio assumed for chondritic meteorites.

MacDonald and Knopoff's (1958) requirement that the mean
atomic number of the core is 23 gives a core of mean composition
$Fe_{86}Si_{14}$ (wt %) if Si is assumed to be the light element. Ringwood's (1966a,b) estimate, which was calculated on the basis of
a pyrolitic mantle and a chondritic earth, gives the Si content
of the whole core as 11 wt %. Balchan and Cowan (1966), using
(Birch, 1965), obtain a Si content of 14-15 wt % on the basis of
compressibility measurements. The above three estimates were
obtained by different methods, and their close agreement is
remarkable. A fluid outer core containing 15-25 wt % Si (or S)
is consistent with the earth models selected by Press (1968).

Ringwood (1966a,b) has suggested that Si in an iron core is
not consistent with a mantle and a crust that contain considerable Fe^{3+}. On the basis of this and a number of other factors he suggests that the core and mantle are in gross chemical
disequilibrium. Ringwood therefore postulates that the deep
interior of the earth accreted cool and that as accretional
energy increased, the temperature increased markedly, so that
iron oxides and some silica were reduced to an Fe-Si alloy and
that vast amounts of CO were outgassed. If the rate of sinking
of metal during core formation was high in comparison to the
rate of chemical reequilibration by diffusion, the core which
separated would not be in equilibrium with the mantle. As the
Fe-Si rich core sinks, the volatile-rich primordial interior is
displaced towards the surface.

It is difficult to imagine how a liquid Fe-Si alloy could
sink through an oxidized mantle and not be at least partially
reoxidized to the point where Si was driven out of the metallic
phase. Brett (1971) points out that equilibration between slag
and metal in blast furnaces occurs in hours to days, so it is
unlikely that extreme disequilibrium could be maintained during core formation, unless masses of metal tens of kilometers
in diameter were involved from the beginning. Brett suggests
that Si may have entered the sinking iron during core formation,
maintaining an equilibrium situation and without requiring the
extreme initial reduction, demanded by Ringwood. He points out
that reactions of the type

$$2(1 - x)Fe + SiO_2 = 2 Fe_{1-x}O + Si$$

or

$$2SiO_2 + 2Fe = Fe_2SiO_4 + Si$$

may favor the right-hand side in the lower mantle, on the basis
of thermodynamic calculations. If this is the case, the problem of extreme core-mantle disequilibrium which appears to be
a problem in Ringwood's model is averted.

Sulfur

Until recently, most geochemists attributed the depletion
in S in rocks at the earth's surface to its high volatility,
which allowed it to be outgassed during and subsequent to
accretion. Recently, a number of investigators, most notably
Murthy and Hall (1970), have revived Goldschmidt's concept of
sulfur deep within the earth by postulating that it is the light
element present in the core. They point out that the extremely
high temperature accretion, massive carbon monoxide blowoff,
and pronounced core-mantle disequilibrium of Ringwood's Si model
appear improbable.

Murthy and Hall (1970) suggest that S is more depleted than
other volatiles including C, N, H_2O, the halogens, and rare
gases in the crust and mantle with respect to ordinary or carbonaceous chondritic meteorites (Figure 1). Reasons for such
preferential volatilization are hard to conceive during or subsequent to accretion, so they suggest that either the fractionation occurred in the nebula prior to accretion or sulfur is
segregated in the interior of the earth. They point out that
relative depletions of volatile elements appear to be more or
less constant from one meteorite type to another and that anomalous depletion of a single element should not be expected.
They therefore suggest that the sulfur segregated with Fe to
form a liquid core of Fe-FeS. Brett and Bell (1969) showed that
the eutectic temperature of the Fe-FeS eutectic at 30 kbar is
virtually unchanged from the eutectic temperature at one atmosphere ($988^\circ C$). Brett and Bell suggested that if S is the light
element in the core, coalescence of core material in the upper
mantle prior to core formation could have taken place at considerably lower temperatures than those required for a pure iron
protocore ($990^\circ C$ versus about $1600^\circ C$). Brett and Bell warn of
the dangers of extrapolating their data to core pressures (over
$1\frac{1}{2}$ orders of magnitude), yet most advocates of a sulfur-rich
core have done so with the use of linear extrapolation. Such
extrapolations have been used to point out that such a core
accreting at relatively low temperature circumvents the thermal
problems presented by a silicon-rich core, which may require
much higher temperatures early in the earth's history. First,
such extrapolations are unwarranted, since melting curves are
certainly not linear, and Taylor and Mao (1970) and King and
Ahrens (1973) have found evidence of a high-pressure polymorph
of FeS, whose presence would change the slope of the pressure-
temperature curve for any composition in the Fe-FeS system.
Second, the Fe-rich Fe-Si eutectic is at $1200^\circ C$ (Hansen and
Anderko, 1958), and at 1 atm, $Fe_{80}Si_{20}$ melts at about $100^\circ C$
lower than $Fe_{80}S_{20}$. No work has been done on the effect of
pressure on Fe-Si melting. Therefore to postulate S instead_

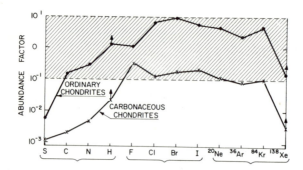

FIGURE 1. Abundances of some volatile elements in the silicate fraction of the earth compared to type 1 carbonaceous
chondrites and ordinary chondrites and a mixture (5:4:1) of
ordinary chondrites, type 1 carbonaceous chondrites, and iron
meteorites. Upward arrows indicate that ratios for H and Xe
represent a minimum (after Murthy (1976), courtesy Wiley,
New York).

of Si as the light element in the core on the basis of arguments
involving low-temperature melting is unwarranted, unless the Si
was incorporated into the metallic Fe after the Fe became molten.

Turekian (in Turekian and Wedepohl, 1961) suggests that the
abundance of S in ultramafic rocks is known only to within an
order of magnitude. Turekian and Wedepohl find that the S content of all sedimentary rocks with the exception of sandstones
is greater than 1000 ppm. These data combined with the data
for basalts reported above surely suggest that the abundance
of S for crust plus mantle (about 150 ppm) reported by Murthy
and Hall (1970) is a low estimate. If Turekian and Wedepohl's
estimate of 300 ppm of S in peridotites is combined with Moore
and Fabbi's data for basalts, a figure of 425 ppm is obtained
for a pyrolitic mantle. Sulfur is thus depleted by about the
same factor as C and N with respect to chondrites but is greatly depleted with respect to the other volatile elements.

Sulfur appears to be a better candidate than Si for the
light element in the core unless the equilibrium alloying of
Si with metallic Fe takes place at high temperatures as suggested by Brett (1971). The reasons given by Murthy and Hall
(1970, 1972) for the presence of S in the core have the following problems associated with them, however;

1. The main argument for S in the core is the depletion of
S in terrestrial rocks relative to chondritic abundances.
Either the earth did not accrete much S-rich material, the S is
in the core, or it remains in the mantle. Moore and Fabbi

(1971) suggest that the best value for the S content of juvenile basalt that has not been subject to extensive outgassing is 800 ± 150 ppm. They suggest that mantle-derived magmas are saturated or near saturated with S. Haughton *et al.* (1974) suggest that the common observation that many layered intrusives have a sulfide-rich layer near their base indicates that the magmas were saturated with S when they were intruded. Anderson (1974), on the basis of analyses of S of glass inclusions in phenocrysts from basalts, suggests that S is universally saturated in magmas prior to eruption. Any S present in the mantle source rock in excess of the saturation limit would form a dense, immiscible Fe-S-O liquid, which has a eutectic temperature of $915^{\circ}C$ (Naldrett, 1969). The high density of this liquid with respect to the silicate melt would surely not allow it to be transported to the surface in appreciable quantities; rather, it would remain in the mantle. Therefore for all three possible models described above, the S content of surface rocks should be low. Hence the depletion of S cannot be used as evidence that S is present in the core.

2. As was pointed out by Ringwood (1966a,b), certain elements, e.g., Ni, Pt, Cu, Co, appear to be enriched in the upper mantle thus being explained. However, all samples used by Murthy and Hall to show S depletion were surely derived from the upper 100-200 km in the earth. Therefore these too were not involved in core formation, so their S abundances cast no light on the presence or absence of S in the core. The argument of Murthy and Hall stands, however, if the outer 100-200 km of the earth was oxidized prior to core formation (J. S. Lewis, personal communication, 1975).

If the material forming the earth accreted at relatively low temperatures (e.g., Lewis, 1974), then the earth should have received the full cosmic abundance of S. If core formation occurred in some manner akin to that suggested by Elasser (1963), it would be remarkable if appreciable S were not present in the core. It seems that S abundances of rocks presently in the crust cannot be used to deduce this, however.

King and Ahrens (1973) on the basis of shock wave experiments have shown that the pressure-density profile for the outer core may be satisfied by an Fe-FeS liquid containing 10-12 wt % S. Usselman (1975) calculates the composition of the core at the core-mantle boundary to be between 13 and 9 wt % S, based on calculations of the densities at high pressures of Fe-FeS liquids.

The presence of appreciable quantities of a light element in the outer core but not in the inner core would certainly lower the liquidus temperature of the outer core with respect to an inner core of near-pure iron. This compositional difference has been used by a number of people (e.g., Urey, 1952)

to explain why the inner core is solid and the outer core is
liquid. The higher pressure of the inner core with respect
to the outer core would increase the solidus temperature and
thus have considerable bearing on the solidity of the inner
core.

We do not even know if temperatures in the core are constant or if convection occurs causing change in temperature at
a given point with time. We do not know if the core is heating
or cooling. If it is cooling, the inner core must be growing
at the expense of the outer core, with solid precipitates sett-
ling down to the inner core-outer core boundary. A multicomponent system in a thermal gradient such as exists in the core
must be layered chemically unless convection and mixing occur.

Potassium in the Core

Lewis (1971) and Hall and Murthy (1971) have made the intriguing speculation that the balk of the earth's potassium is
present in the earth's outer core, which they assume to be S-
rich. If this is true, the implications are immense, since
^{40}K is the main source of radioactive heat within the earth.
If the heat source were in the core and thus very deep, energy
would be available for core convection, thus providing the
convective mechanism probably necessary for the earth's magnetic field. Energy would also be available for convection
in the mantle, thus providing the energy for sea floor spreading and continental drift.

The major arguments used by the above authors and Goettel
and Lewis (1973) are that thermodynamic calculations predict
that reactions of the type $K_2O + H_2S = K_2S + H_2O$ proceed from
left to right at elevated temperatures. They also point out
that potassium-rich sulfides occur in some meteorites and that
some analyses of Mansfield slag show that K is concentrated in
the sulfide phase. Geottel (1972) reports significant addition
of K into the sulfide phase in experiments in which molten Fe-
FeS was heated with roedderite ($K_2Mg_5Si_{12}O_{30}$) at 988° and
1030°C. The partitioning strongly favors the silicate phase,
but even so, a considerable amount of K could be partioned into
the core. Goettle (1975) reports significant K partitioning
in Fe-FeS melt (195 and 290 ppm of K at 1030° and 1070°C, respectively) in equilibrium with potassium feldspar. Little
(about 10 ppm) K was observed in the Fe-FeS melt equilibrated
with the Forest City chondrite at the same temperature, however. Potassium and the other alkali metals are considerably
depleted in the upper mantle with respect to chondrites (Gast,
1960). The conventional explanation (Gast, 1960) is that being
volatile elements, they never condensed because of the high

temperatures pertaining during condensation. Proponents of K in the core argue that the K was condensed but was partioned strongly into the core during core formation. One of the major arguments for placing K in the core is therefore analogous to an important argument for placing S in the core.

Oversby and Ringwood (1972) state that Lewis (1971) and Goettel and Lewis (1973) base their thermodynamic arguments exclusively on oxide-sulfide reactions and that silicate assemblages, especially those involving aluminosilicates, are more realistic for the conditions in the earth and tend to stabilize K in oxidized form. They suggest that experiments involving roedderite are irrelevant, since this mineral is not representative of the mantle. They present data on the partitioning of K between molten basalt and Fe-FeS liquids which they believe indicate that an upper limit of about 1 wt % of the earth's available K would be partitioned in the core. Goettel and Lewis (1973) claim that if one were to accept Oversby and Ringwood's conclusion, even this small amount is sufficient to promote convection in core and mantle and that experiments involving molten basalt are not relevant to the problem. Oversby and Ringwood (1973) further claim that neither the metallurgical data nor the occurrences of K-rich sulfides in meteorites have any bearing on K partitioning between core and mantle.

To sum up, the possible presence of K in the core is an intriguing speculation. However, there are insufficient experimental data to confirm or deny the possibility. Carefully designed experiments involving measurements of the partitioning of K between a series of molten sulfide and solid silicate compositions over a range of temperatures, pressures, and oxygen fugacities are required to settle this issue.

Other Elements

It is logical to assume that elements other than Fe, Ni, S, and/or Si must be present in the core in trace and minor amounts. Those elements that have greater affinity for metallic iron than for silicates or oxides are likely to be present in the core in appreciable amounts if they are cosmochemically abundant and if the core formed from an initially homogeneous earth.

Brett (1971) estimated the abundance of a number of elements in the mantle and then calculated the abundances of these elements in the core, assuming a chondritic earth and partitioning ratios between metal and silicate from meteoritic data. Thermodynamic data predicted that the same nonvolatile elements that are depleted in the mantle should be enriched in the core. Brett states that it is therefore likely that a few tenths of 1% of Mn, Cr, and P occur in the core. One might also expect

a few tenths of 1% (Murthy and Hall, 1970) to several percent
C to also be present, the amount depending on the extent of
early outgassing in the earth and the temperature of accretion.

If S is present in the core, the chalcophile elements
should be enriched in the core with respect to the mantle.
Ringwood (1966a,b) has pointed out that several of these elements are depleted in the upper mantle with respect to chondrites.

CORE-MANTLE EQUILIBRIUM

Ringwood (e.g., 1966a,b) in advocating Si as the light element in the core pointed out that a Si-rich core indicates
chemical disequilibrium between the core and mantle, as was
discussed earlier. Ringwood states that additional evidence
for core-mantle disequilibrium is (1) the apparent presence of
considerable Fe^{3+} in the mantle, which can hardly be in chemical equilibrium with metallic iron, (2) the relatively high
abundance of the elements, Ni, Co, Cu, Pt, Pd, Ir, Au, and Os
in the mantle, since these elements strongly partitioned into
the metallic phase rather than the silicate phase, and (3) the
gases being emitted from volcanoes are enriched in CO_2 and
H_2O and therefore cannot be in chemical equilibrium with a core
of metallic iron.

Turekian and Clark (1969) point out that in their model in
which the metallic core is accreted first, one would not expect
chemical equilibrium between core and mantle, since the bulk of
the core and mantle scarcely communicate with one another
chemically.

Brett (1971) suggests that the observations of Ringwood can
be explained in terms of an equilibrium model. Thermodynamic
calculations, involving large extrapolations, indicate that
Si should be partitioned into the metal phase at the expense of
the silicate phase at depth and the reverse should occur for
Ni. Convection may have brought up Ni-rich silicates to the
upper mantle. Brett points out that there is evidence suggesting that much of the mantle may be in equilibrium with metallic
Fe and that portions of the upper mantle have been oxidized by
contamination by the crust and the reactions with gases. Escape of hydrogen causes oxidation of lavas (Sata, 1972). Mao
(1974), on the basis of high-pressure experiments, states that
FeO decomposes to Fe_3O_4 + Fe or Fe_2O_3 + Fe under conditions of
the lower mantle. The apparent disequilibrium of Fe^{3+} in the
upper mantle and Fe^0 in the core could thus be explained.
There is a suggestion from the data of French (1966) that H_2O
dominates over H_2 for metallic Fe-silicate melt-gas systems at
high pressures.

Ringwood's (1966a,b,1971) objections to core-mantle equilibrium on the basis of gases emitted from volcanoes therefore appear to be unwarranted. The fact that CO_2 and H_2O are the major H and C gaseous species is consistent with the relatively high abundances of carbonates and H_2O in the crust of the earth. Gases emitted from blast furnaces are not in equilibrium with the chemical regime at their place of evolution.

Recent experimental work on partitioning of Au and Re between iron and silicates has shown that the upper mantle and crust appear to be out of chemical equilibrium with the core with respect to these elements (Kimura et al., 1974). The problem of equilibrium versus disequilibrium remains unsolved, but it surely is critical to understanding the processes of core formation. The major problem lies in extrapolating thermodynamic data to lower mantle pressures and temperatures: it is a tempting game but may be little more than that.

CONCLUSIONS

At present we have a reasonably good idea of the pressure regime in the core and fairly broad limits of the amount of light element required in the outer core. We totally lack solid evidence on what that light element might be. Reasonable agreement exists among investigators that the core-forming process occurred early, but we lack knowledge of that process apart from the vaguest outline. The effect of core formation on the distribution of elements in the mantle is unknown. The nature of the chemical reactions that surely must occur at the core-mantle interface remains a mystery. The temperature regime within the core is known only imprecisely. Clearly, an answer to these and other questions would greatly improve our knowledge of the early history of the earth.

The following geochemical studies would help refine our knowledge of the core.

1. Further research is needed on the partitioning of trace elements in crustal and mantle rocks and between silicates and metallic iron in order to determine whether or not trace elemental abundances can teach us anything about abundances of elements in the core.

2. Extension of the pressures at which static high-pressure phase equilibria can be performed is needed, so that melting temperatures in the core can be better delineated and the earth's thermal regime can be better understood. Research on the factors governing the extrapolation of high-pressure data to higher pressures would also be of great importance.

3. Further high-temperature/high-pressure research is required to test the possibility of K going into a sulfide phase at ehe expense of silicates. Similarly, the suggestion that silica plus iron goes to silicon dissolved in iron plus iron oxide at temperatures pertaining to the lower mantle should be tested experimentally.

4. Further measurements on the electrical conductivities and seismic properties of Fe-S and Fe-Si alloys should be made in the hope of finding one pair that better fits the properties of the core than the other.

ACKNOWLEDGMENTS

I thank J. S. Lewis, T. R. McGetchin, V. Rama Murthy, and Motoaki Sato for helpful reviews.

REFERENCES

Alder, B. J. (1966). Is the mantle soluble in the core? *J. Geophys. Res. 71*, 4973-4979.

Al'tshuler, L. V. (1971). Composition and state of matter in the deep interior of the earth, *Phys. Earth Planet. Interiors 5*, 295-300.

Al'tshuler, L. V., A. A. Bakanava, and R. F. Trunin (1962). Shock adiabats and zero isotherms of seven metals at high pressure. *Sov. Phys. JETP* (Engl. Transl.) *15*, 65-74.

Anderson, A. T., Jr. (1974). Chlorine, sulfur and water in magmas and oceans, *Geol. Soc. Amer. Bull. 85*, 1485-1492.

Anderson, D. L., and T. C. Hanks (1972). Formation of the earth's core, *Nature 237*, 387-388.

Anderson, D. L., C. Sammis, and T. Jordan (1971). Composition and evolution of the mantle and core, *Science 171*, 1003-1112.

Balchan, A. S., and G. R. Cowan (1966). Shock compression of two iron silicon alloys to 2.7 megabars, *J. Geophys. Res. 71*, 3577-3288.

Birch, F., (1952). Elasticity and composition of the earth's interior. *J. Geophys. Res. 57*, 227-286.

Birch, F., (1963). Some geophysical applications of high pressure research, *in "Solids Under Pressure,"* W. Paul and S. M. Warschauer, ed., McGraw-Hill, New York.

Birch, F., (1964). Density and composition of mantle and core, *J. Geophys. Res. 69*, 4377-4388.

Birch, F., (1965). Speculations on the earth's thermal history, *Geol. Soc. Amer. Bull. 76*, 133-154.

Birch, F., (1972). The melting relations of iron and temperatures in the earth's core, *Geophys. J. 29*, 373-387.

Bird, J. M., and M. S. Weathers, (1975). Josephinite; Specimens from the earth's core?, *Earth Planet. Sci. Lett. 28*, 51-64.

Brett, R., (1971). The earth's core: Speculations on its chemical equilibrium with the mantle, *Geochim. Cosmochim. Acta 35*, 203-225.

Brett, R., and P. M. Bell, (1969). Melting relations in the Fe-rich portion of the Fe-FeS system at 30 kb pressure, *Earth Planet. Sci. Lett. 6*, 479-452.

Buchbinder, G. G. R., (1972). An estimate of inner core density. *Phys. Earth Planet. Interiors 5*, 123-128.

Bullen, K. E., (1936). The variation of density and ellipticities of strata of equal density within the earth, *Mon. Notices Roy. Astron. Soc., Geophys. Suppl. 3*, 395-401.

Bullen, K. E., (1973). Cores of the terrestrial planets, *Nature 243*, 68-70.

Clark, S. P., Jr., (1963). Variation of density in the earth and the melting curve in the mantle, *in "The Earth Sciences"*, T. W. Donnelly, ed., University of Chicago Press, Chicago, Ill.

Clark, S. P., and A. E. Ringwood, (1964). Density distribution and constitution of the mantle, *Revs. Geophys. Space Phys. 2*, 35-88.

Clark, S. P., K. K. Turekian, and L. Grossman, (1972). Model for the early history of the earth, *in "The Nature of the Solid Earth*, E. C. Robertson, ed., pp. 3-18, McGraw-Hill, New York.

Dana, J. D., (1875). *"Manual of Geology,"* 2nd ed., Ivison, Blakeman, Taylor, New York.

Elsasser, W. M., (1963). Early history of the earth, *in "Earth Science and Meteoritics,"* Geiss and E. D. Goldberg, ed., pp. 1-30, North-Holland, Amsterdam.

Eucken, A., (1944). Physikalisch-chemisch Betrachtungen über die fruheste Entwicklungsgeschichte der *Erde, Nachr. Akad. Wiss. Goettingen, Math. Phys. Kl. 1*, 1-25.

Flaser, F. M., and F. Birch, (1973). Energetics of core formation: A correction, *J. Geophys. Res. 78*, 6101-6103.

French, B. M., (1966). Some geological implications of equilibrium between graphite and a C-H-O gas phase at high temperatures and pressures, *Rev. Geophys. Space Phys. 4*, 223-253.

Gast, P. W., (1960). Limitations on the composition of the upper mantle, *J. Geophys. Res. 65*, 1287-1297.

Goettel, K. A., (1972). Partitioning of potassium between silicate and sulfide melts; experiments relevant to the earth's core. *Earth Planet. Sci. Lett. 6*, 161-166.

Goettel, K. A., (1975). Potassium in the earth's core. Ph.D. thesis, 136 pp., Mass. Inst. of Technol., Cambridge, Mass.

Goettel, K. A., and J. S. Lewis, (1973). Comments on a paper by V. M. Oversby and R. E. Ringwood, *Earth Planet. Sci. Lett. 18,* 148-150.

Grossman, L., (1972). Condensation in the primitive solar nebula, *Geochim. Cosmochim. Acta 36,* 597-619.

Grossman, L. A., and S. P. Clark, Jr., (1973). High temperature condensation in chondrites and the environment in which they formed, *Geochim. Cosmochim. Acta 37,* 635-650.

Grossman, L. A., and J. W. Larimer, (1974). Early chemical history of the solar system, *Revs. Geophys. Space Phys. 12,* 71-101.

Hall, H. T., and V. R. Murthy, (1971). The early chemical history of the earth: Some critical elemental fractionations, *Earth Planet. Sci. Lett. 11,* 239-244.

Hanks, T., and D. L. Anderson, (1969). The early thermal history of the earth, *Phys. Earth Planet. Interiors 2,* 19-29.

Hansen, M., and K. Anderko, (1958). *"Constitution of Binary Alloys,"* 2nd ed., McGraw-Hill, New York.

Haughton, D. R., P. L. Roeder, and B. J. Skinner, (1974). Solubility of sulfur in mafic magmas, *Econ. Geol. 69,* 451-462.

Higgins, G., and G. C. Kennedy, (1971). The adiabatic gradient and the melting point gradient in the core of the earth. *J. Geophys. Res. 76,* 1870-1878.

Kimura, K., R. S. Lewis, and E. Anders, (1974). Distribution of gold and rhenium between nickel-iron and silicate melts: Implications for the abundance of siderophile elements on the earth and moon, *Geochim. Cosmochim. Acta 38,* 683-702.

King, D. A., and T. J. Ahrens, (1973). Shock compression of iron sulfide and the possible sulfur content of the earth's core, *Nature Phys. Sci. 293,* 82-84.

Kraut, E. A., and G. C. Kennedy, (1966). New melting law at high pressures, *Phys. Rev. 151,* 668-675.

Lewis, J. S., (1971). Consequences of the presence of sulfur in the core of the earth, *Earth Planet. Sci. Lett. 11,* 130-134.

Lewis, J. S., (1973). Chemistry of the planets, *Annu. Rev. Phys. Chem. 24,* 339-351.

Lewis, J. S., (1974). The chemistry of the solar system, *Sci. Amer., 230(3)* 50.

Lyttleton, R. A., (1973). The end of the iron-core age, *Moon 7,* 422-439.

Mao, H. K., (1974). A discussion of the iron oxides at high pressure with implications for the chemical and thermal evolution of the earth, *Carnegie Inst. Wash. Yearb. 73,* 511-518.

MacDonald, G. J. F., and L. Knopoff, (1958). The chemical composition of the outer core, *J. Geophys. 1*, 284-297.

Mason, B., (1966). Composition of the earth, *Nature 211*, 616-618.

McQueen, R. G., and S. P. Marsh, (1966). Shock wave compression of iron-nickel alloys and the earth's core. *J. Geophys. Res. 71*, 1751-1756.

Moore, J. G., and B. P. Fabbi, (1971). An estimate of the juvinile sulfur content of basalt, *Contrib. Mineral. Petrol. 33*, 118-127.

Murthy, V. R., (1976). Composition of the core and the early chemical history of the earth, *in "Early History of the Earth."* B. F. Windley, ed., Wiley, New York.

Murthy, V. R., and H. T. Hall (1970). The chemical composition of the earth core: Possibility of sulfur in the core, *Phys. Earth Planet. Interiors 2*, 276-282.

Murthy, V. R., and H. T. Hall (1972). The origin and composition of the earth's core, *Phys. Earth Planet. Interiors 6*, 123-130.

Naldrett, A. J., (1969). A portion of the system Fe-S-O between 900° and 1808°C and its application to sulfide ore magmas, *J. Petrology 10*, 171-201.

Oldham, R. D., (1906). The constitution of the interior of the earth as revealed by earthquakes, *Quart. J. Geol. Soc. London 62*, 456-475.

Oversby, V. M., and A. E. Ringwood (1971). The time of formation of the earth's core, *Nature 234*, 463-465.

Oversby, V. M., and A. E. Ringwood (1972). Potassium distribution between metal and silicate and its bearing on the occurrence of potassium in the earth's core, *Earth Planet. Sci. Lett. 14*, 345-347.

Oversby, V. M., and A. E. Ringwood (1973). Reply to comments by K. A. Goettel and J. S. Lewis, *Earth Planet. Sci. Lett. 18*, 151-152.

Press, F., (1968). Density distribution in earth, *Science 160*, 1218-1221.

Ramsey, W. H., (1948). On the constitution of the terrestrial planets, *Mon. Notices Roy. Astron. Soc. 108*, 406-413.

Ringwood, A. E. (1958). Constitution of the mantle, 3, Consequences of the olvine-spinel transition, *Geochim. Cosmochim. Acta 15*, 195-212.

Ringwood, A. E. (1959). On the chemical evolution and densities of the planets, *Geochim. Cosmochim. Acta 15*, 257-283.

Ringwood, A. E., (1960). Some aspects of the thermal evolution of the earth, *Geochim. Cosmochim. Acta 20*, 241-259.

Ringwood, A. E., (1961). Silicon in the metal phase of enstatite chondrites and some geochemical implications, *Geochim. Cosmochim. Acta 30*, 41-104.

Ringwood, A. E., (1966a). Chemical evolution of the terrestrial planets, *Geochim. Cosmochim. Acta 30*, 41-104.

Ringwood, A. E. (1966b). The chemical composition and origin of the earth, *in "Advances in Earth Science,"* P. M. Hurley, ed., MIT Press, Cambridge, Mass.

Ringwood, A. E., (1971). Core and mantle equilibrium: Comments on a paper by R. Brett, *Geochim. Cosmochim. Acta 35*, 223-229.

Ringwood, A. E. (1975). *"Composition and Petrology of the Earth's Mantle,"* McGraw-Hill, New York.

Sato, M., (1972). Intrinsic oxygen fugacities of iron-bearing oxide and silicate minerals under low total pressure, *Geol. Soc. Amer. Mem. 135*, 289-307.

Sterrett, K. F., W. Klement, Jr., and G. C. Kennedy, (1965). Effect of pressure on the melting of iron. *J. Geophys. Res. 70*, 1979-1984.

Stewart, R. M. (1973). Composition and temperature of the outer core, *J. Geophys. Res. 78*, 2586-2597.

Stull, D. R., et a., (1971). *JANAF Thermochemical Tables, Nat. Ref. Data Ser.*, Vol. 37, U. S. National Bureau of Standards, Washington, D. C.

Taylor, L. A., and H. K. Mao, (1970). A high-pressure polymorph of troilite, FeS, *Science 170*, 850-851.

Taylor, S. R., and P. Jakes, (1974). The geochemical evolution of the moon, *Geochim. Cosmochim. Acta, Suppl. 5*, 1287-1305.

Toksöz, M. N., A. M. Dainty, S. C. Solomon, and K. R. Anderson, (1973). Velocity, structure, and evolution of the moon, *Geochim. Cosmochim. Acta, Suppl. 4*, 2529-2547.

Turekian, K. K., and S. P. Clark, Jr., (1969). Inhomogeneous accumulation of the earth from the primitive solar nebula, *Earth Planet. Sci. Lett. 6*, 346-348.

Turekian, K. K., and K. H. Wedepohl, (1961). Distribution of the elements in some major units of the earth's crust, *Geol. Soc. Amer. Bull. 72*, 175-192.

Urey, H. C., (1952). *"The Planets"*, Yale University Press, New Haven, Conn.

Urey, H. C. (1960). On the chemical evolution and densities of the planets, *Geochim. Cosmochim. Acta 18*, 151-153.

Usselman, T. M., (1975). Experimental approach to the state of the core, 2, Composition and thermal regime, *Amer. J. Sci. 275*, 291-303.

Weichert, E., (1897). Ueber die Massenvertheilung in Innern der Erde, *Nachr. Ges. Wiss. Goettingen, Math. Phys. Kl. 3.*

4

COMPARATIVE PLANETOLOGY
AND THE ORIGIN OF CONTINENTS

Paul D. Lowman, Jr.

Geophysics Branch, Goddard Space Flight Center, Greenbelt, Maryland

*This chapter summarizes and extends an earlier study of com-
parative crustal evolution in terrestrial planets. It is shown
that the Moon, Mercury, and Mars underwent a common pattern of
early crustal evolution after a high-temperature origin: formation of a global differentiated crust, a period of impact, and
a period of basaltic magmatism. Mars continued into a fourth
stage of crustal uplift and incipient rifting, but stopped evolving 1 to 3 billion years ago. The terrestrial Precambrian record appears consistent with a similar evolutionary pattern up
to about 2.5 billion years ago, when plate tectonic processes
began. The early crust of the Earth was of global extent and
intermediate bulk composition. It has been repeatedly redifferentiated by partial melting and remobilization, forming a
vertically zoned continental crust with a granulitic lower part
and a granitic upper part. The Earth's continental crust is, in
this theory, primarily the result of very early global differentiation; the present continents are the redifferentiated remnants
of the original intermediate crust.*

I. INTRODUCTION

The exploration of space, which has greatly stimulated the
study of "comparative planetology" (a term first coined by
George Gamow (1948)), is beginning to illuminate fundamental
problems of terrestrial geology, in particular, the origin and
early evolution of continents. The main reason for this is that
the geologic record on the Moon and the terrestrial planets is

strongest where that of the Earth is weakest, namely, for the
period corresponding to the early Precambrian (or Archean)
(Cloud, 1971). And since there is good terrestrial evidence
that most continental growth took place in the Precambrian, the
results of lunar and planetary exploration have major implications for the origin of the continents. The purpose of this
paper is to summarize those implications, to discuss differences
between terrestrial and extraterrestrial crustal evolution, and
to describe some tectonic features and their relation to the
petrologic evolution of continental crust.

The range of subject matter that must be covered in a treatment of this sort is immense. This chapter is a condensation
and further development of an earlier paper in the *Journal of
Geology* (Lowman, 1976a), to which the reader is referred for a
more detailed discussion and list of references. To keep this
chapter within practical limits, only those references not included in the bibliography of the earlier paper will be listed
here.

PATTERNS OF CRUSTAL EVOLUTION IN SILICATE PLANETS

The silicate or terrestrial planets, including the Moon,
cover a wide range of mass, chemical composition, atmospheres,
and surficial geology. But they all appear to have had generally
similar early evolutionary patterns (Fig. 1), with the possible
exception of Venus, for which the data are still fragmentary.
These patterns have been systematized as follows.

Stage I: Origin and Heating

This stage is largely the province of astrophysics and cos-
mochemistry, and hence will not be discussed in detail. However, a few main aspects of the origin of silicate planets in
general are especially relevant to crustal evolution and to the
origin of continents.

First, and most important, is the nearly conclusive evidence
that the Moon and by implication the terrestrial planets began
their geologic histories at very high temperatures. Although
the initial stages of accretion may have been at low temperatures, as proposed some 25 years ago by Harold Urey, various
processes raised the temperatures of these bodies greatly, at
least locally to the melting point, by the time they had attained
their present sizes (Fricker *et al.*, 1973). This is demonstrated
clearly by the abundant evidence for high temperatures and very
early differentiation in the Moon, the smallest member of the inne[r]
solar system except for Phobos and Deimos. Further support for

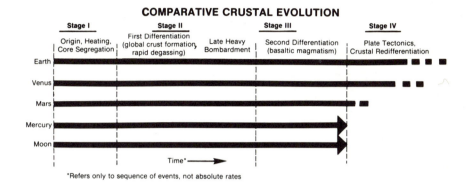

FIGURE 1. Schematic outline of crustal evolution in silicate planets.

early high temperatures comes from the asteroids, generally agreed to be the parent bodies for meteorites. Chapman (1976) has presented strong arguments that more than 100 of these bodies, in the 200 to 500 km range, were melted and differentiated within about 200 million years after the solar system itself was formed.

It was shown, before the Apollo lunar missions, by Safronov (1969) and Hanks and Anderson (1969), that the Earth's formation should have been accompanied by high temperatures due to energy of accretion and core segregation. When these studies are added to the evidence for intense heating in bodies smaller than the Earth, it becomes quite clear that the Earth itself must have been very hot at the beginning of geologic history some 4.6 billion years ago. It is generally agreed that such conditions would, *a priori*, favor early and extensive differentiation.

A second aspect of the origin of planets that should be mentioned here is the possibility of primordial layering, due to heterogeneous accretion processes, which has been suggested by Turekian and Clark (1969) and others to be at least partly responsible for the concentric planetary structure that others attribute to differentiation. Recent studies by Hartman (1976), based on the work of Wetherill, have provided another mechanism

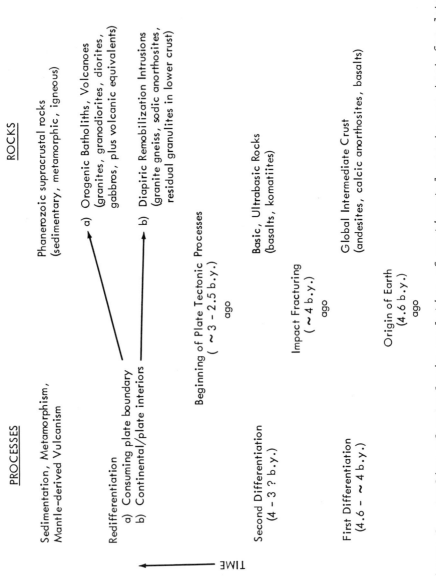

Figure 2. Outline of petrologic evolution of continental crust; see text for details and supporting evidence, and also Table I.

TABLE I. Crustal Evolution of the Earth

Stage (time before present)	Events and Processes	Terrestrial evidence
I (4.6 b.y.)	(a) Origin by rapid, homogeneous accretion.	(a) Age of Earth from Pb isotope data; homogeneous accretion inferred from magmatic character of terrestrial crustal compositon.
	(b) Rapid heating, leading to extensive melting of outer part of Earth, degassing, and core formation.	(b) Thermal studies; formation of core before oldest known rocks (3.8 b.y.); concentration of U, Th, rare earths in crust; loss of rare gases.
II (4.6–4 b.y.)	(a) First differentiation, by igneous processes: large-scale magmatic differentiation of molten mantle or partial melting of mantle or partial melting of mantle with subsequent magmatic differentiation. Resultant global crust had intermediate bulk composition equivalent to andesite; actual rock types included diorite, andesite, basalt, high-Ca anorthosite.	(a) Early differentiation indicated by Pb isotope data; magmatic origin of early crust indicated by geochemical studies of distribution coefficients; bulk composition inferred from present average composition of continental crust and geophysical evidence for intermediate lower crust. Rock types inferred from surviving Archean assemblages and petrologic studies. Global nature of original crust suggested by analogy with Moon; little direct terrestrial evidence.

TABLE I (Continued)

Stage (time before present)	Events and Processes	Terrestrial evidence
II	(b) Impact fracturing of early crust, overturning early layering and resetting radiometric dates, peaking at 4 b.y. ago with formation of large basins covering 2/3 of the Earth's surface.	(b) Existence of large though relatively young (1-2 b.y.) impact features in shield areas demonstrates infall of large bodies. Impact history of Moon suggests higher flux earlier; closeness to Earth requires similar terrestrial flux history.
III (4 - 2.5 b.y.)	(a) Second differentiation, by generation of basic and ultrabasic magmas under Stage II impact basins by pressure release and fracturing. Result was short-lived ensialic ocean basins.	(a) Archean greenstone belts over 2.5 b.y. old with basalts and periodotitic pillow lavas; field relations indicate deposition on preexisting crust, later remobilized.
	(b) Uplift and rifting of ensialic ocean basins leading to beginning of sea-floor spreading and other plate tectonic processes. Number of plates ("microplates") very large and tectonic evolution rapid.	(a) Thermal calculations indicate weakening of crust, localization of magma generation in impact basins. Small size and large number of Archean plates indicated by geometry of greenstone belts.

Stage (time before present)	Events and Processes	Terrestrial evidence
III	(c) Redifferentiation by partial melting of primitive crust began, leading to zoned continental crust (granites above, granulites below). Minor continuing evolution of sial from mantle.	(b) Redifferentiation indicated by experimental and theoretical petrology, field relations in granite/greenstone terranes, and evidence for vertical zonation in existing continental crust.
IV (2.5 b.y. to present)	(a) Coalescence of microplates, crustal thickening, and stabilization.	(a) Coherent regional structural patterns in shield areas; great increase in abundance of surviving rocks from this time.
	(b) Beginning of plate tectonic processes equivalent to those of present, with relatively small number of major plates and long plate boundaries.	(b) Increase in K/Na ratios of rocks, relative abundance of monzonites, and maturity of sediments after 2.5 b.y. ago, indicating major change in tectonic style at about this time.

for such externally produced layering. Hartmann has shown that considerable fractions of the material added to the inner planets in the later stages of accretion could have originated in planetesimals elsewhere in the solar systems with different compositions resulting from condensation sequences (Lewis, 1972).

These studies thus suggest that the silicate planets were initially layered to some extent. However, as pointed out in Lowman (1976a), it seems clear that the chemistry of terrestrial continental crust and of the lunar highland rocks, which, except for the achondrites, are the closest available analogs to a primordial planetary crust, is fundamentally of an igneous nature, suggesting homogeneous accretion or homogenization shortly after accretion. The following discussion will therefore assume such an origin.

Stage II: First Differentiation

This stage is an extremely important one in its implications for the origin of terrestrial continents. One of the most interesting results of lunar and planetary exploration has been the discovery that the Moon and probably Mars, Mercury, and Venus underwent extensive early differentiation, forming global igneous crusts. Since the even smaller meteorite parent bodies, with diameters less than 500 km, were also differentiated by melting between 4.4 and 4.6 billion years ago (Short, 1975), it now seems clear that such differentiation is the rule for most silicate planets.

Of the primitive planetary crusts formed by this process, that of the Moon, the lunar highlands, is the best known. This crust has, however, been intensely cratered, brecciated, and locally remelted, and a simple petrographis description is difficult. Chemically, it is equivalent to a mixture of 70% anorthositic gabbro and 30% high-aluminum basalt (Taylor, 1974). The main processes that formed the lunar highland crust, which I have grouped as the "first differentiation" (Lowman, 1972), have been elegantly reconstructed from chemical, mineralogical, and petrologic evidence by Taylor (1975). They can be summarized as follows. The Moon was, in Taylor's theory, extensively melted, at least in its outer parts, during its formation, thus constituting in essence a large magma body or "magma ocean" (Wood, 1975). This magma underwent differentiation, largely by fractional crystallization, on cooling, resulting in an initially layered structure consisting from the surface down of (1) a chilled undifferentiated outer crust, (2) a plagioclase-rich layer, and (3) an olivine-orthopyroxene layer. Trace elements and lithophile elements (especially potassium) accumulated in residual liquid in the plagioclase-rich zone. These processes were largely completed by about 4.5 billion years ago.

Heavy impact bombardment rapidly disrupted this original structure, producing deep brecciation, mixing, and overturning. The residual liquids described above intruded the crust, forming KREEP basalts. Mixing by impact of the plagioclase-rich layer, the surface crust (about 10 km thick), and the residual liquids produced the anorthositic gabbro that dominates the present highland crust (Adler *et al.*, 1972).

The first differentiation is thus seen to have been, in the Moon, a complex and violent period in which the effects of heavy postaccretion impact were superimposed on a zoned structure produced by magmatic processes. Continuing igneous activity in the 4.5 to 4 b.y. period complicated the lunar geologic record still further. This stage can be considered to have ended with the "lunar cataclysm" described by Tera and Wasserburg (1974), the short catastrophic period of major impacts that formed the circular mare basins about 4 b.y. ago, resetting most highland radiometric dates. Mars and Mercury also underwent a period of relatively late bombardment, producing basins such as Hellas (Mars) and Caloris (Mercury) whose stratigraphic ages are at least consistent with a comparable age. When added to the evidence for cratering (presumably impact) on Venus and Ganymede (Frieden and Swindell, 1976), it appears clear that a discrete pulse of heavy bombardment hit all the planets of the inner solar system after the initial postaccretion impacts.

Stage III: Second Differentiation

The next major event in the evolution of the Moon, Mercury, and Mars was the formation of large areas of dark, level terrain: the maria on the Moon or "smooth plains" on Mercury. On the Moon and Mars, at least, this terrain is a series of basic to ultrabasic lava flows and possibly intrusives. This basic magnatism is generally agreed to have been a process of partial melting, localized by impact, of the residuum from the first differentiation, and will therefore be referred to as the "second differentiation."

On the Moon, this period lasted nearly a billion years, from about 4 to 3 billion years ago, although there was probably later minor volcanism in areas such as the Marius Hills and Tsiolkovsky. At its end, the Moon and other planets had a twofold physiographic dichotomy superficially analogous to the continent-ocean basin dichotomy of the Earth. However, the lunar maria are floored by highland crust, and thus are not truly similar to ocean basins.

The second differentiation was the last major internal event in the crustal evolution of the Moon and Mercury. Mars, however, presumably because of its greater mass and consequently greater internal energy, evolved somewhat further before "winding down" some 2.5 billion years ago (Neukum and Wise, 1976; Malin, 1976).

Stage IV: Tectonic Fracturing, Plate Tectonics, Redifferentiation

The spectacular Valles Marineris of Mars, although furrowed with dentritic tributaries of probably fluvial origin, is almost certainly a tectonic depression, and is generally compared to the Red Sea-African Rift Valley system. It thus appears to represent initial tectonic fracturing that if continued would lead to sea-floor spreading and related plate tectonic processes. However, it seems unlikely that Mars has moved into a true plate tectonic stage; in addition to the physiography of the Valles Marineris, the great size of volcanic piles such as Olympus Mons implies that the Martian crust has been stationary relative to the presumed mantle plumes responsible for them (Carr, 1976).

Recent earth-based radar studies of Venus (Goldstein *et al.*, 1976; Campbell *et al.*, 1976) reveal irregular, noncircular features of regional size that may be tectonic basins. Venus appears to have a hybrid physiography somewhat like that of Mars, with both cratered terrain and tectonic depressions, as might be expected from its greater mass. However, there is nothing comparable on Mercury or the Moon. Of the silicate planets, only the Earth has moved fully into stage IV, and a discussion of this stage will therefore be postponed until the next section.

EARLY CRUSTAL EVOLUTION IN THE EARTH

At first glance, the Earth would appear utterly different from the other bodies of the inner solar system, and is, of course, geologically unique in many ways. However, when the incomplete and complex evidence from Precambrian rocks is examined in the light of the crustal evolution patterns just discussed, it is possible to trace a broadly similar sequence of events and processes (Fig. 2, Table I). This sequence, which has been used in Lowman (1976a) as the basis for a new theory for the origin of continents, will be only briefly summarized here.

Stage I: Origin and Heating

This stage will not be discussed in detail. However, it should be pointed out that evidence for high temperatures in the formation of the Moon virtually requires high temperatures in the formation of the Earth, at least in the later stages, because of the greater mass and lower surface-to-volume ratio of the latter. As mentioned previously, Safronov (1969) and Hanks and Anderson (1969) have independently deduced such conditions, as have, more recently, Shaw (1971) and Condie (1976).

Stage II: First Differentiation

A large body of terrestrial geologic evidence, summarized earlier (Lowman, 1976a), indicates that the Earth's continental crust was developed largely in early Precambrian time, and it was proposed that this process was the terrestrial first differentiation. Since that paper was published, similar conclusions have been reached by several authorities. A number of these were presented at the 1975 NATO conference on the early history of the Earth (Windley, 1976); of particular interest are those of D. M. Shaw, J. V. Smith, and B. Jahn, and L. E. Nyquist, all of whom proposed or implied early differentiation of the Earth. Siever (1974) and Condie (1976) have also recently proposed early, rapid crustal growth. However, details of the process remain obscure, to say the least.

If homogeneous accretion of the Earth is assumed, for reasons discussed earlier, there would be two general paths by which the first differentiation could occur. If the Earth was formed solid, or solidified immediately after forming, magma generation by partial melting of the mantle would probably be the dominant process, as it is now. On the other hand, a largely or wholly molten Earth would be in essence an immense magma chamber, and formation of an early crust would depend largely on various processes of magmatic differentiation such as fractional crystallization (complicated, of course, by the effects of high pressure at great depths).

In the theory proposed in Lowman (1976a), I suggested that generation of andesitic magmas by partial melting under very high water pressure was the dominant means by which the early crust was formed. Two problems arise with this sequence of events. First, it seems unlikely that plate tectonic processes like those of today operated earlier than about 2.5 billion years ago (Burke and Dewey, 1973), and present-day andesites seem firmly tied to subduction (e.g., Marsh, 1976). Second, andesites or their metamorphic equivalents seem to be largely absent from "lower Archean" terranes (Barker and Arth, 1976), which tend to consist of bimodal trondhjemite-basalt suites. These problems require more study; however, tentative answers to them can be at least outlined.

On the question of whether andesites can be formed without plate tectonic processes, the work of Fanale (1971) is of interest. Fanale inferred "catastrophic early degassing" and concurrent differentiation for the primitive Earth on the Basis of rare gas studies. This event, in which water must have been a major component, might have produced suitable conditions for partial melting of the primitive mantle, as shown by the experiments of Lambert and Wyllie (1970). This pulse of andesite formation could have been a unique event, after which, as proposed by Holmes (1965), the mantle "would have lost forever its

capacity for generating andesite," later andesites being the result of recycling in what are now called subduction zones. The
"lower Archean" terranes described by Barker and Arth (1976)
might have formed in the interval between the primordial andesite
pulse and the much later onset of plate tectonics with associated
andesites. This sequence of events might also help solve the
problem of low initial $^{87}Sr/^{86}Sr$ ratios discussed by Moorbath
(1975) in that the primordial andesites would have formed from
an undifferentiated mantle with primitive strontium and a low
Rb content.

It appears then that my original model for the first differentiation is still a possible one. However, it is not clear
whether it is the *actual* one. As we have seen, the structure
and lithology of the Moon seem best explained by extensive early
melting and consequent fractionation of the "magma ocean," and
it is hard to see how the early Earth could have avoided comparable melting in view of its greater mass, lower surface-to-volume
ratio, and higher volatile content. Therefore, we must examine
the possibility of the second of the two evolutionary paths
mentioned earlier, large-scale magmatic processes in a largely
molten Earth. This is, of course, a very old concept that has
been generally discarded in recent decades since it was demonstrated that the mantle is solid and that most magmas developed by
partial melting of the mantle or crust.

D. M. Shaw has studied in great detail (1972, 1976) the petrologic evolution of a molten Earth. His work, especially the
earlier paper, is of particular interest in that it was based
almost entirely on terrestrial geochemical evidence, thus providing an independent line of reasoning with which the models
based on comparative planetology can be compared.

Shaw pointed out that three major processes probably operated in the primitive Earth: fractional crystallization, gravi-
tation-controlled diffusion, and, in solic portions, zone melting.
He proposed (1972) that these processes produced, in the first
few hundred million years of the Earth's history, "all the materials of the present continental crust," forming a global primordial crust 14 km thick, with the bulk chemical composition of
quartz-diorite. He further proposed that there had been little
crustal growth since that time. It will be obvious that Shaw's
model, of which I was unaware, is fundamentally similar to the
one I proposed independently in 1976.

In his more recent paper, Shaw has modified this model somewhat, suggesting a protocrust consisting of a global 14 km layer,
similar to the present continental crust, overlaid by basic rock.
He has further suggested that heavy impact would have disrupted,
mixed, and remelted this crust, a proposal similar to that of
Taylor (1975) for impact mixing of the early lunar crust.

Further studies of the analogy between lunar and terrestrial crustal evolution are clearly called for. At this time, only a few general conclusions about the development of the early Earth will be presented, as follows.

First, high temperatures must be assumed in any study of the evolution of the primordial crust. Second, it seems highly probable that the Earth did undergo a first differentiation analogous to that of the Moon, producing a global igneous crust. General considerations, in particular the greater volatile and alkali content of the Earth, dictate that this crust would have been more sialic than that of the Moon (Hargraves, 1976). At this time, it is not possible to make a firm choice between the solid and molten Earth models, but it appears that either could yield a global crust of intermediate bulk composition, whether by, respectively, partial melting under high water pressures or by fractionation of a molten mantle. I originally proposed a petrographic description of this crust: "a complexly interlayered series of andesites and subordinate basalts, locally intruded by coeval diorites and granodiorites" with "layered feldspar-rich intrusives with local occurrences of very calcic anorthosite." This model still seems fundamentally valid, although, following Taylor and Shaw, the importance of pre-4 b.y. impact mixing and overturning should be pointed out. These processes probably contributed to widespread resetting of radiometric ages, thus helping to answer the long-standing question of why no truly primitive crust has been identified.

Early differentiation can be inferred for the Earth with some confidence even without reference to extraterrestrial geology. However, it is not possible to so infer a *global* differentiation, as opposed to localized formation of sialic nuclei, other than by reasoning that the energetic, volatile-rich Earth should have undergone differentiation at least as extensive as that in the smaller planets. But the general outlines of the processes by which the originally global terrestrial crust was disrupted can be inferred (Frey, 1976, and this volume).

The first event in these processes was almost certainly a period of heavy impact bombardment of the Earth analogous to the basin-forming impacts on the Moon and other terrestrial planets (Schmitt, 1975). There are no known terrestrial impact structures even close to 4 b.y. old, but it is impossible to imagine how the Earth could have escaped such a bombardment (Murray *et al.*, 1975). This event can be placed by analogy at the end of stage II. It has been suggested by Goodwin (1974) and others that these impacts might have initiated igneous events leading to the formation of sialic nuclei, but this seems unlikely for reasons discussed in Lowman (1976a).

Stage III: Second Differentiation and Permobile Tectonism

There is abundant evidence for a long period of widespread terrestrial basic and ultrabasic magmatism whose products survive in the Archean greenstone belts found in all shield areas. The abundance, relative age, and composition of these rocks suggest that this episode was analogous to the formation of the lunar maria and, more speculatively, the marelike smooth plains of Mars and Mercury. They have in fact been termed "terrestrial maria" by Green (1972).

This stage has been the subject of an intensive theoretical investigation by Frey (1976), who has studied in particular the thermal effects of the presumed 4 b.y. impact period. He has shown that the results of such impacts would be pressure release and steepening of the thermal gradients under the basins so formed. These effects would lead to partial melting of the mantle and generation of basaltic magma. Of equal interest, however, is Frey's conclusion that these conditions would increase mantle convection at these sites, thus leading to uplift, rifting, and spreading of these primitive ocean basins. This would clearly be analogous to present-day plate tectonic processes. However, since there would be, according to Frey's calculations, many impact basins, covering about half the area of the Earth, we should expect many small plates, or microplates, to form. This situation has been deduced, on purely terrestrial evidence, by Burke *et al.* (1976), who propose a "permobile" stage characterized by rapid lateral crustal movement. They do not consider this stage as plate tectonics.

Frey's concept of the transition from stage II to stage III is much more satisfactory than the one I proposed in 1976, involving crustal foundering as a means of basin formation. Furthermore, to the extent that the processes of this permobile stage resemble those of present plate tectonics, his proposal may account for the degree of similarity between Archean igneous rocks and those of the plate tectonic stage, a topic discussed by Windley (1976b), who concluded that some sort of plate activity did operate in the Archean. Details of stage III are difficult to unravel because ocean basins tend to destroy the evidence of their birth. However, there are now known to be many failed rifts, or aulocogens (Burke and Dewey, 1973a), as old as 2.2 billion years, that may be similar to those with which true tectonic ocean basins began (Burke *et al.*, 1976).

Stage IV: Plate Tectonics

There was probably no sudden end to stage III. Nevertheless, it is now generally agreed that the Earth's evolution underwent a major change in style about 2.5 billion years ago. The work of Engel *et al.* (1974), among others, indicates that several petrochemical indices, such as the K_2O/Na_2O ratio in rocks, turned toward their present values at about that time. Since plate tectonic processes (including attendant igneous activity) can account for many characteristics of Phanerozoic petrology, at least along plate margins, it appears likely that the plate tectonic stage began about 2.5 billion years ago. Furthermore, for reasons discussed in Lowman (1976a), it is believed that the ocean basins and continents had attained roughly their present volume by that time. There are, of course, no major occurrences of oceanic crust this old, but this is presumably due to recycling; the oceanic crust can thus be much younger than the ocean basins.

In this theory, the continents are considered the greatly altered remnants of an originally global differentiated crust, rather than concentric accretions of orogenic belts or coalesced sialic nuclei, and ocean basin growth rather than continental growth the dominant mode of crustal evolution for the last four billion years. When first proposed (Lowman, 1972, 1973), the theory differed sharply from most current opinion. However, since that time there has been a major shift of opinion toward the concept of a very early, widespread crust, as shown by several papers presented at the 1975 NATO conference mentioned earlier (Windley, 1976a) and publications by Hargraves (1976), Burke *et al.* (1976), and Condie (1976).

Crustal evolution obviously continued after 2.5 billion years ago; let us now consider one aspect of this evolution.

REDIFFERENTIATION OF THE PRIMITIVE CRUST

On the basis of reasoning discussed in Lowman (1976a) the primitive global crust of the Earth was inferred to have an intermediate bulk composition, and to have consisted specifically of andesites, basalts, diorites, granodiorites, and anorthosites. This description does not fit the presently surviving Archean crust as a whole, except for local occurrences. In particular, the exposed shield areas are dominated by granites and granitic gneisses. To explain this discrepancy, it was proposed that the primitive crust had undergone "redifferentiation" by partial melting, remobilization, and diapiric intrusion, leaving a granulitic residuum now making up the lower continental crust. Several lines of evidence not explored in detail in the earlier paper support this suggestion; they can be summarized as follows.

1. Field Evidence

Many studies of deeply exposed Archean terranes indicate that granulite-facies rocks are a major part of the lower crust (Goldsmith, 1976). The work of Lambert (1971) in Australia is especially important on this point. He showed that in two nearby shield areas, the deeper rocks were chiefly granulites and the shallower ones chiefly granites. He proposed that the latter had been formed by partial melting of the original crust, leaving the granulites as a residuum. (It may be of interest that I was unaware of Lambert's work when I published the same idea in 1976.) Field studies in other areas, described by Sighinolfi (1972), Hyndman (1972), Read and Watson (1975), and Burke et al. (1976) similarly suggest the importance of granulites in the lower crust. Many inclusions brought up from the lower crust in diatremes are granulitic; McGetchin and Silver (1971) describe such inclusions from the Colorado Plateau, although these were, of mafic and not intermediate composition.

2. Geophysical Evidence

Laboratory measurements of compressional and shear wave velocities in granulites at pressures up to 10 kbar have been made by Christensen and Fountain (1974), and are similar to those measured for the lower continental crust. They concluded that granulites are a major constituent of the lower crust. Heat flow measurements in the Australian shield reported by Lambert (1971) indicate that the lower crust must be depleted in radioactive elements, also suggesting granulites (specifically of intermediate composition, in Lambert's view). Gravity data in the Kalgoorlie area are consistent with this interpretation.

3. Petrologic Evidence

Experimental evidence, discussed by Lambert and Wyllie (1970) and Fyfe (1973), when viewed in context with field and geophysical evidence, strongly suggests that the lower continental crust must partially melt through geologic time in a variety of tectonic environments. Burke and Dewey (1973b) and Sighinolfi (1971) arrived at the same conclusion, suggesting that the first-formed liquid would be granitic and the residuum granulitic. Fyfe further demonstrated that this process could produce diapiric

Figure 3. Gemini 6 photograph of diapiric granite gneiss domes in the Yetti plain (Dorsale Reguibat) of North Africa. Linear features at center are seif dunes; spacecraft nose at lower right. See Lowman (1972a) for discussion.

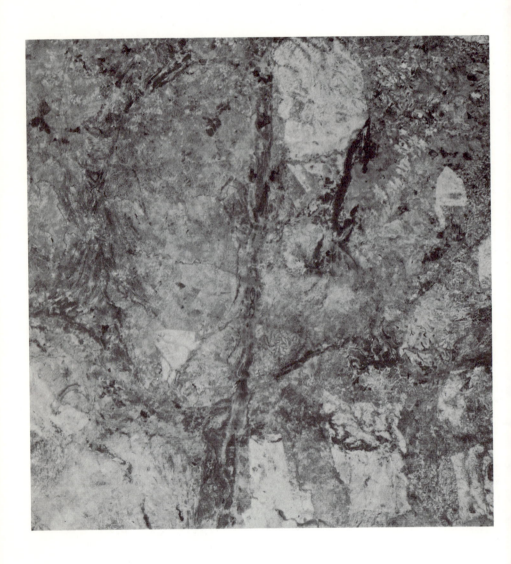

Figure 4. Landsat-1 image of Rhodesia, showing Great Dike
(center) and adjacent gneiss domes ("gregarious batholiths"),
visible as large, elliptical masses. Bright angular patches
at right (east of Great Dike) are overgrazed areas, not bedrock
features. (Image 1103-07285, 3 Nov. 72.)

Figure 5. Landsat-1 image of Pilbara district, northwestern Australia, showing diapiric granite gneiss domes (large, light-colored masses). See Lowman (1976b) for discussion.

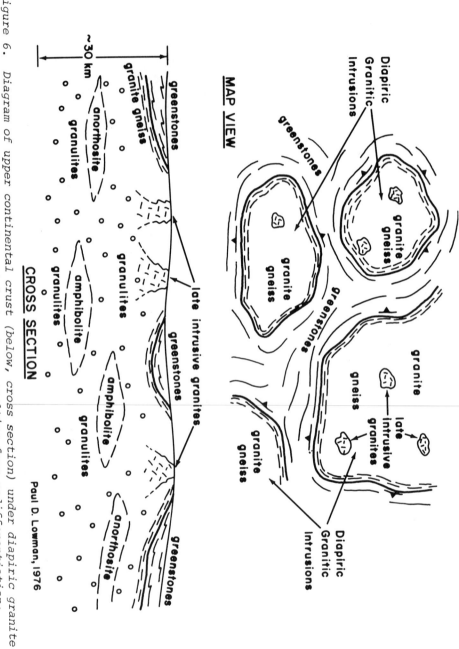

MAP VIEW

Diapiric
Granitic
Intrusions

greenstones

greenstones

granite
gneiss

granite
gneiss

granite
gneiss

greenstones

granite
gneiss

granite
gneiss

granite

late
intrusive
granites

Diapiric
Granitic
Intrusions

CROSS SECTION

~30 km

granite gneiss

greenstones

anorthosite

granulites

granulites

amphibolite

granulites

late intrusive granites

greenstones

granulites

amphibolite

granulites

anorthosite

greenstones

Paul D. Lowman, 1976

Figure 6. Diagram of upper continental crust (below, cross section) under diapiric granite gneiss domes, showing vertically zoned crustal structure resulting from redifferentiation: granulites and related rock types below, granites above, intruding greenstone belts.

granite intrusions such as the well-known "gregarious batholiths" of Rhodesia described by MacGregor (1951) as well as small igneous intrusions.

The suggestion of Burke and Dewey, that the Tibetan Plateau has undergone remelting to produce silicic magmas, has been strikingly confirmed by the recent discovery of Hamet and Allegre (1976), who have reported an initial $^{87}Sr/^{86}Sr$ ratio of above 0.740 for the Oligocene Manaslu granite of central Nepal.

4. Orbital Photography

Orbital photography of well-exposed Precambrian shield areas strongly suggests that diapiric granite intrusions are much more widespread than generally realized (Lowman, 1972a). Examples are presented from Africa and Australia (Figs. 3-6). Furthermore, detailed geologic maps of shield areas in other regions, showing the predominance of granitic rocks (especially granitic gneisses), indicate that such diapiric granites are much more common than igneous granites (orgenic batholiths?). If such areas as the Dorsale Reguibat of North Africa give a representative exposure of the lower crust, it may be suggested that similar diapiric intrusions underly continental crust anywhere that they have not been destroyed by later igneous events. These features may account for many of the vague arcuate features seen on orbital photographs (Short et al., 1976), perhaps as the result of slight tectonic movements along their contacts. Some of those features, such as the Vredefort Dome, may be the result of intrusion at the site of former impact craters; to that extent, such impacts would promote redifferentiation.

The proposed mechanism for redifferentiation of the primitive intermediate crust is clearly supported by a wide range of evidence, and by much work not cited at length, such as that of Fahrig and Eade (1968) and Burwash and Krupicka (1970) in the Canadian shield. However, it is a greatly simplified model in many ways; for example, most Precambrian terranes show evidence of several periods of igneous and metamorphic activity, rather than the simple sequence described here. Furthermore, the silicic igneous rocks of Archean areas generally have low radiogenic Sr contents (Moorbath, 1975), apparently contradicting derivation from a preexisting sialic crust. Perhaps this results from still earlier metamorphism that depleted the supposed preexisting sialic crust in Rb (Hargraves, 1976), or from recycling in the permobile stage as proposed by Armstrong (1968) for present-day subduction zones.

A topic requiring further study is the question of whether appreciable Phanerozoic redifferentiation of sialic crust has taken place at consuming plate boundaries over subduction zones, a problem closely related to that of the source of granitic mag-

mas.   Three lines of evidence support this possibility. First,
many orogenic batholith rocks have intermediate initial Sr ratios
(Faure and Powell, 1972), possibly indicating partial derivation
of their magmas by melting of sialic crust. Second, petrologic
evidence from the Sierra Nevada batholith (Presnall and Bateman,
1973) indicates that andesitic magmas rising from a subduction
zone could have had enough superheat to partially melt sialic
crust they intruded. Field relations in the Southern California
batholith described by Gastil (1975) also suggest derivation of
at least the eastern part of the batholith by partial fusion
near ascending plutons. It seems likely that the result of
such processes in general would be to generate magmas more silicic than the crust from which they were formed. However, the
actual result would be greatly complicated by magmatic differentiation, assimilation, and other factors, and it is thus not certain that something equivalent to the proposed redifferentiation
of anorogenic regions would in fact happen.

Recent analyses of Martian soil performed (Clark *et al.*, 1976)
by the X-ray fluorescence experiment on Viking landers 1 and 2
are relevant to the redifferentiation process proposed here. As
discussed by the Viking XRF team (Baird *et al.*, 1976), the four
analyses indicate very low potassium content, from which they
concluded that there are no large, exposed, alkali-rich granite
occurrences on Mars. More generally, they inferred that the XRF
results indicate a "poorly differentiated" planet, in apparent
contradiction to the conclusion of the Mariner 9 IRIS investigators (Hanel *et al.*, 1972).

There is no reason to doubt the Viking team's conclusion that
there is little granite exposed on Mars. However, this does not
necessarily rule out widespread stage II planetary differentiation. As discussed here and in Lowman (1976a) granites on the
Earth are essentially secondary rock types, resulting largely
from partial melting ("redifferentiation") of preexisiting crust.
The primordial crust probably did not contain much granite (Fig.
2), which in this concept results from long-continued tectonic
and petrologic activity. Since it seems clear that Mars stopped
evolving internally at least one billion years ago, after just
entering Stage IV, one would not expect abundant granite. The
Viking XRF team's conclusion that there apparently had not been
on Mars "large-scale planetary differentiation of a terrestrial
nature" may be quite correct in that terrestrial differentiation has involved much remelting and mobilization of the Earth's
original global crust.

Similar reasoning can explain why granites are so scarce on
the Moon. However, the prevailing depletion of volatiles in the
Moon must be added to the lack of continuing geologic activity
and redifferentiation. Lunar magmas would be low in alkalis to
begin with, and being dry, would tend to differentiate along a
Skaergaard trend with only minor silicic end products (Walter,
1965).

SUMMARY

In this chapter I have tried to show that despite their many
differences in mass, volatile content, and bulk composition, the
terrestrial planets seem to have evolved along similar lines
during their early history. The major difference among them is
how far they evolved. Small bodies such as the Moon essentially
stopped developing at a relatively early stage, about 3 b.y. ago
in the case of the Moon. Mars had more heat initially and cooled
more slowly because of its greater mass, but it too apparently
stopped 2 - 3 b.y. ago. The Earth, most massive of the terrestrial planets, is a boiling cauldron of geologic activity by
comparison, whose crustal evolution has gone far beyond that of
the other planets and shows little sign of slowing down.
On the basis of this common early evolutionary pattern, and
terrestrial geologic evidence summarized in Lowman (1976a), I
have proposed that the Earth's continents are the greatly altered
remnants of an originally global primitive crust. Terrestrial
crustal evolution in this theory has been dominated for the last
three billion years or more by formation of ocean basins, rather
than by continental growth. Most of the continental crust
formed early in the Earth's history; apparent formation of new
sial has in fact been the result of recycling of old sial through
subduction zones or crustal remelting in anorogenic areas. It
remains to be seen whether apparently contradictory evidence
from strontium isotopes can be convincingly reconciled with this
concept.
It has been further proposed that the original crust of the
Earth has been repeatedly redifferentiated by partial melting
and mobilization of large-ion lithophile elements. The result
of this process has been a vertically zoned crust, with granulites (largely of intermediate composition) below, and granites,
granite gneisses, and migmatites above, overlaid of course in
most areas by younger sedimentary and volcanic rocks. This
redifferentiation process, to which several independent lines
of evidence point, provides at least part of the answer to the
question of why granites are so common on the Earth but so scarce
on the Moon and Mars. Most terrestrial granites are the result
of secondary processes dependent on continued internal activity,
both within plates and at plate boundaries. A planet that stops
evolving with the stage III basaltic flooding or early in stage
IV will probably not develop large amounts of granite beyond the
minor amounts formed by classical magmatic differentiation in
igneous intrusions.
The theory presented here will doubtless require revision as
new knowledge becomes available. But even in its present form,
it offers a coherent framework for interpreting a wide range of
geologic evidence from the planets of the inner solar system,
and for understanding the early evolution of the Earth. In addition, the theory has some general implications for future developments in planetology.

It has been realized for many years (Cameron, 1968) that the well-known Hertsprung-Russell (H-R) diagram is also a stellar evolutionary chart with which, in principle, the evolution of any star can be predicted if its initial mass and composition are known. This diagram was derived, of course, from observations of tens of thousands of stars, in contrast to the few objects available to students of planetary evolution. Nevertheless, the outlines of a planetary equivalent to the H-R diagram are beginning to take shape. With reference to crustal evolution in silicate planets, the following generalizations can be suggested.

Stellar evolution is governed in principle by only two factors: mass and initial composition (Hoyle, 1955). Silicate planet postaccretionary evolution is similarly governed by mass and composition, but also by impact history. Of these three factors, mass appears to be by far the dominant one. This relationship is at least suggested, though of course not proved, by the geology of Mercury and the Moon. These two bodies are fairly close in mass, Mercury having about 5% and the Moon about 1% the mass of the Earth, but differ considerably in composition, judging from their densities. Yet there is little apparent difference in their early evolutionary patterns.

Despite the speculative nature of this generalization, it may be suggested that a planetary "H-R diagram" showing degree of crustal evolution as a function of mass (with allowance for composition and impact history) is an achievable goal for planetologists. To construct such a diagram, further information is badly needed in the following areas: physiography and surface composition of Venus; surface composition (especially highlands) of Mercure; surface composition (especially highlands) of Mars; physiography and composition of the large Jovian and Saturnian satellites; compositions of the asteroids; and resolution of the tektite problem.

ACKNOWLEDGMENTS

I am indebted to Herbert Frey for many stimulating discussions of the problems of ocean basin formation. W. J. Hinze and M. A. Mayhew brought several useful papers on redifferentiation to my attention.

REFERENCES[1]

1. Baird, A. K., Toulmin, P., III, Clark, B. C., Rose, H. J.,
    Jr., Keil, K., Christian, R. P., and Gooding, J. L., (1976).
    Mineralogic and petrologic implications of Viking geochemi-
    cal results from Mars: interim report, *Science 194*, 1288-
    1293.
2. Barker, F., and Arth, J. C., (1976). Generation of trondh-
    jemitic-tonalitic liquids and Archean bimodal trondhjemite-
    basalt suites, *Geology 4*, 596-600.
3. Burke, K. C. A., and Dewey, J. F. (1973a). An outline of
    Precambrian plate development, *in* Implications of Continen-
    tal Drift to the Earth Sciences, Vol 2 (D. H. Tarling and
    S. K. Runcorn, eds., pp. 1035-1045. Academic Press, New
    York.
4. Burke, K., Dewey, J. F., and Kidd, W. S. F. (1976). Domi-
    nance of horizontal movements, arc and microcontinental
    collisions during the later permobile regime, *in* "The Early
    History of the Earth" (B. F. Windley, ed.), pp. 113-129.
    Wiley (Interscience), New York.
5. Burbash, R. A., and Krupicka, J. (1969). Cratonic reacti-
    vation in the Precambrian basement of western Canada. I.
    Deformation and chemistry, *Canad. J. Earth Sci. 6.*, 1381-1396.
6. Cameron, R. C. (1968). Stellar evolution *in* "Introduction
    to Space Science," 2nd ed. (W. N. Hess and G. D. Mead, eds.),
    pp. 873-936. Gordon & Breach, New York.
7. Campbell, D. B., Dyce, R. B., and Pettengill, G. H. (1976).
    New radar image of Venus, *Science 193*, 1123-1124.
8. Carr, M. H. (1976). The volcanoes of Mars, *Sci. Am. 234*,
    32-43.
9. Chapman, C. R. (1976). Asteroids as meteorite parent-bodies:
    The astronomical perspective, *Geochim. Cosmochim. Acta
    40*, 701-719.
10. Christensen, N. I., and Fountain, D. M. (1975). Constitu-
    tion of the lower continental crust based on experimental
    studies of seismic velocities in granulite, *Bull. Geol.
    Soc. Am. 86*, 227-236.
11. Clark, B. C., Baird, A. K., Rose, H. J., Jr., Toulmin, P.,
    III, Keil, K., Castro, A. J., Kelliher, W. C., Rowe, C. D.,
    and Evans, P. H. (1976). Inorganic analyses of Martian
    surface samples at the Viking landing sites, *Science 194*,
    1283-1288.

[1]*References cited in text but not listed here will be found
in the bibliography of Lowman (1976a).*

12. Cloud, P. E. (1970). Lunar science and planetary history, *Science, 169,* 1114 (editorial), 18 Sept. 19 .

13. Condie, K. C. (1976). "Plate Tectonics and Crustal Evolution." Pergamon, New York. 288

14. Dewey, J. F., and Burke, K. C. A. (1973). Tibetan, Variscan, and Precambrian basement reactivation: Products of continental collision, *J. Geol. 81,* 683-692.

15. Fahrig, W. R., and Eade, K. E. (1968). The chemical evolution of the Canadian Shield, *Canad. J. Earth Sci. 5,* 1247-1252.

16. Fanale, F. (1971). A case for catastrophic early degassing of the earth, *Chem. Geol. 8,* 79-105.

17. Frey, H. (1977). Crustal evolution in the early Earth: Basin-forming impacts, crustal dichotomy, and plate tectonics, Ph.D. Thesis, Dept. of Astronomy, Univ. of Maryland, College Park.

18. Fricker, P. E., Reynolds, R. T., and Summers, A. L. (1974). On the thermal evolution of the terrestrial planets, *The Moon 9* 211-218.

19. Frieden, B. R., and Swindell, W. (1976). Restored pictures of Ganymede, Moon of Jupiter, *Science 191,* 1237-1241.

20. Fyfe, W. S. (1973). The generation of batholiths, *Tectonophysics 17,* 273-283.

21. Gamow, G. (1948). "Biography of the Earth." *New American Library,* New York.

22. Gastil, R. G. (1975). Plutonic zones in the Peninsular Ranges of southern California and northern Baja California, *Geology 3,* 361-363.

23. Goldsmith, J. R. (1976). Scapolites, granulites, and volatiles in the lower crust, *Bull. Geol. Soc. Am. 87,* 161-168.

24. Goldstein, R. M., Green, R. R., and Rumsey, H. C. (1976). Venus radar images, *J. Geophys. Res. 81,* 4807-4817.

25. Hamat, J., and Allegre, C. (1976). Rb-Sr systematics in granite from central Nepal (Manaslu): significance of the Ologocene age and high $^{87}Sr/^{86}Sr$ ratio in Himalayan orogeny, *Geology, 4,* 470-472.

26. Hargraves, R. B. (1976). Precambrian geologic history, *Science 193,* 363-371.

27. Hartmann, W. K. (1976). Planet formation: compositional mixing and lunar compositional anomalies, *Icarus 27,* 553-559.

28. Holmes, A (1965. "Principles of Physical Geology," Ronald Press, New York.

29. Hyndman, D. W. (1972), "Petrology of Igneous and Metamorphic Rocks. "McGraw-Hill, New York.

30. Lambert, I. B. (1971), The composition and evolution of the deep continental crust, *Spec. Pub. Geol. Australia 3,* 419-428.

31. Lambert, I. B., and Wyllie, P. J. (1970. Melting in the deep crust and upper mantle and the nature of the low velocity layer, *Phys. Earth Planet. Interiors 3*, 316-322.

32. Lewis, J. S. (1972). Low-temperature condensation from the solar nebula, *Icarus 16*, 241-252.

33. Lowman, P. D., Jr. (1972). "The Third Planet." Weltflugbild, Zurich.

34. Lowman, P. D., Jr. (1972b). The geologic evolution of the Moon, *J. Geol.*, *80* 125-166.

35. Lowman, P. D., Jr. (1973). Evolution of the Earth's crust: Evidence from comparative planetology, X-644-73-322, Goddard Space Flight Center, Greenbelt, Maryland.

36. Lowman, P. D., Jr. (1976a). Crustal evolution in silicate planets: Implications for the origin of continents, *J. Geol. 84*, 1-26.

37. Lowman, P. D., Sr. (1976b). A satellite view of diapiric Archean granites in Western Australia, *J. Geol. 84*, 237-238.

38. MacGregor, A. M. (1951). Some milestones in the Precambrian of southern Rhodesia, *Trans. Geol. Soc. South Africa 54*, 27-71.

39. Marsh, B. D. (1976). Some Aleutian andesites: their nature and source, *J. Geol. 84*, 27-45.

40. Malin, M. C. (1976). Age of Martian channels, *J. Geophys. Res. 81*, 4825-4845.

41. McGetchin, T. R., and Silver, L. T. (1972). A crustal-upper mantle model for the Colorado Plateau based on observations of crystalline rock fragments in the Moses Rock dike, *J. Geophys. Res. 77*, 7022-7037.

42. Neukum, G., and Wise, D. U. (1976). Mars: A standard crater curve and possible new time scale, *Science 194*, 1381-1387.

43. Presnall, D. C., and Bateman, P. (1973). Fusion relations in the system Ab-An-Or, $Q-H_2O$, and generation of granitic magmas in the Sierra Nevada batholith, *Bull. Geol. Soc. Am. 84*, 3181-3202.

44. Read, H. H., and Watson, J. (1975). "Introduction to Geology", V. 2. Macmillan, London.

45. Safronov, V. S. (1969). Evolution of the protoplanetary cloud and formation of the Earth and the planets, *Izdat. Nauka*, NASA Technical Trans. TT F-677 (1972).

46. Schmitt, H. H. (1975). Evolution of the Moon: The 1974 model, *Space Sci. Rev. 18*, 259-279.

47. Shaw, D. M. (1972). Development of the early continental crust. Part 1. Use of trace element distribution coefficient models for the protoarchean crust, *Canad. J. Earth Sci. 9*, 1577-1595.

48. Shaw, D. M. (1976). Development of the early continental crust. Part 2: Prearchean, Protoarchean, and later eras, *in* "The Early History of the Earth" (B.F. Windley, ed.), pp. 33-53. Wiley (Interscience), New York.

49. Short, N. M., Lowman, P. D., Jr., Freden, S. C., and Finch, W. A. (1976). Mission to Earth: Landsat Views the World, NASA SP-360, Government Printing Office, Washington, D.C.

50. Siever, R. (1974). Comparison of Earth and Mars as differentiated planets, Icarus 22, 312-324.

51. Sighinolfi, G. P. (1971). Investigations into deep crustal levels: Fractionating effects and geochemical trends related to high-grade metamorphism, *Geochim. Cosmochim. Acta 35*, 1005-1021.

52. Taylor, S. R. (1975). "Lunar Science: A Post-Apollo View." Pergamon, New York.

53. Walter, L. S. (1965). Lunar differentiation processes, *Ann. N. Y. Acad. Sci. 123*, 470-480.

54. Windley, B. F. (1976). "The Early History of the Earth." Wiley (Interscience), New York.

55. Wood, J. A. (1974). Lunar petrogenesis in a well-stirred magma ocean, Abstracts of Papers Submitted to *Lunar Sci. Conf. 6th, Pt.II, pp. 881-883. Lunar Science Institute, Houston, Texas.*

5

ORIGIN OF THE EARTH'S OCEAN BASINS: IMPLICATIONS FOR THE DEVELOPMENT OF EXTRATERRESTRIAL LIFE[1]

Herbert Frey

Astronomy Program, University of Maryland, College Park, Maryland

If the Earth began its evolution with a global, low-density crust, then some process other than plate tectonics must be responsible for the formation of the original oceanic/continental crustal division. Furthermore, this crustal dichotomy must predate the onset of plate tectonic processes, which require the existence of subductable crust. The lunar crustal dichotomy can be traced back 4 billion years to the period of intense basin-forming impact cratering and the subsequent filling of these basins by liquid basalts. Scaling up from the observed number of lunar basins - a lower limit - to the greater gravitational capture cross section of the Earth and the greater impact velocity on the Earth suggests that more than 50% of the Earth's original global crust was converted into low-elevation basin topography 4 billion years ago. Flooding of these basins by basaltic magmas occured on a time scale short compared to the isostatic adjustment of the basin itself. The formation of these original, (lunar mare-type) ocean basins was followed by the onset of plate tectonic events that completely reworked the oceanic part of the Earth's crust. The 4 billion year old basin-forming epoch, which seems to have been common to the entire inner solar system, may not be a common event in other planetary systems. Large oceans may be absent on otherwise "terrestrial" planets, and this will significantly affect the development of organisms in these environments.

[1]*This work was supported by NASA grant number NGL 21-002-033 and is from a dissertation submitted in partial fulfillment of the requirements for a Ph.D. in Astronomy from the University of Maryland.*

INTRODUCTION

The existence and evolution of life are dependent on the physical environment in which it develops. This dependence is not well known in terms of the origin of life, but it is clear that very different environments will encourage significant differences in subsequent development. The precise nature of the physical environment in which life on this planet began is not completely known, but it is thought that early oceans played an important role. The origin of ocean basins, like the origin of continents, is a matter of controversy among geologists, but it is a question of fundamental importance if the Earth is to be considered a basis for discussing the origin of life on other worlds. This chapter addresses the question of the origin of the Earth's original ocean basins.

Recently Lowman (1976) has suggested that there is a common pattern in the evolution of the smaller terrestrial planets, which also applies to the early Earth. He argues that the Earth began its evolution with a global, low-density crust; the present continents are the much reworked, redifferentiated remnants of this crust. This scenario requires conversion of the global crust into the modern two fold division of ~40% low-density continental crust and ~60% higher density oceanic crust; that is, it becomes necessary to explain the origin of the basic dichotomy in crustal materials, and, in the context of an original global continental crust, to explain the origin of the ocean basins.

Modern oceanic crust is created by seafloor spreading at midocean ridges as part of the plate tectonic evolution of the Earth. This newly formed (high-density) crust is compensated by corresponding destruction of older crust at subduction zones. McKenzie (1973) has shown that continental crust more than 5 km thick is too buoyant to be subducted; once formed, continental crust is not easily destroyed. Subduction destroys only older *oceanic* crust. There seems to be no way, through the processes of modern plate tectonics, to convert more than 50% of an original crust of low-density continental material into the modern crustal dichotomy. These processes can generate the higher density crust of the ocean basins but are not able to eliminate the required volume of low-density material into which the higher density crust is injected.

Furthermore these same plate tectonic processes cannot occur unless the crustal dichotomy is already established because subductable crust is required for subduction to compensate for newly formed oceanic crust. Engel *et al.* (1974) infer from a variety of geochemical parameters that a major change in tectonic style occurred on the Earth 2.5 billion years ago,

perhaps marking the onset of modern (large-plate) plate tectonics. Wise (1974) argues that the modern ratio of oceans/
continents has persisted throughout most of geologic time.
From the above, it would appear that the crustal dichotomy of
the Earth was established early in its history.
    The Earth's crustal dichotomy *superficially* resembles the
high-density-low-density mare-highland division of the Moon,
and Mercury, (and, to a degree Mars). The *lunar* crustal dichotomy is a consequence of a period of intense cratering by
basin-forming objects some 4 billion years ago (Tera *et al.*,
1974). These basins were subsequently flooded by high-density
basaltic liquids erupting from below. This basin-forming event
and its corresponding crustal destruction may have been a common event in the inner solar system (Murray *et al.* , 1975;
Chapman, 1976). Wetherill (1975, 1976) has shown that dynamically plausible orbits existed for such an occurence. In any
case it is not possible for the Earth to have avoided the flux
of objects responsible for forming the lunar mare basins 4
billion years ago. This chapter therefore addresses the question of whether or not this period of crustal disruption was
quantitatively adequate to convert more than 50% of an original
global low-density crust into the modern crustal dichotomy.

TERRESTRIAL IMPACT PARAMETERS

    A lower limit to the number of basin-forming impacts that
must have occurred on the Earth 4 billion years ago can be obtained by scaling upwards from the *observed* number of lunar basins to the greater capture cross section and impact velocity
of the Earth. Because the Earth and Moon at that time were in
roughly their present orbital configurations, both bodies should
have experienced the same spatial distribution of incoming objects. It is therefore reasonable to treat this as a scaling
problem.
    Consider a group of objects deflected into Apollo-type
(Earth-crossing) orbits. These asteroids represent the closest
modern example of the basin-forming objects that impacted the
Moon 4 billion years ago (Wetherill, 1975). Such objects will
approach the Earth-Moon system with a relative velocity between 15 and 20 km/sec (Öpik, 1966). Figure 1a shows the impact
velocity at the surfaces of the Earth and Moon, and the ratio
of these impact velocities, as a function of approach velocity.
The 15-20 km/sec range of approach velocities is shown by the
bar. The impact velocity at the Earth varies from 18.7 to 22.9
km/sec; for the Moon, the corresponding range is 15.2-20.1 km/
sec. The ratio of impact velocities is 1.23 to 1.14. Equiva-

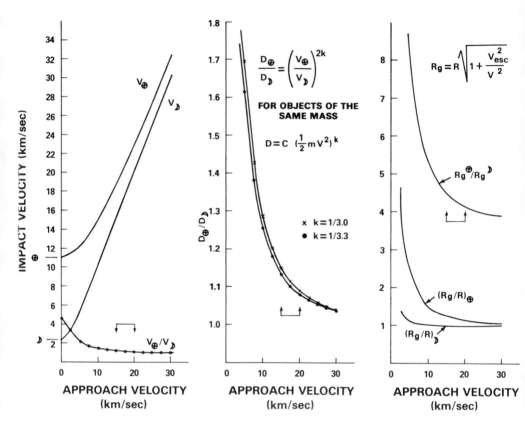

FIGURE 1. Impact parameters as a function of approach velocity. (a) Impact velocity at the surface of the Earth (V⊕)
and the Moon (V☽) and the ratio of these. (b) Ratio of basin
diameters on the Earth to those on the Moon (for objects of
identical mass) for two different values of k. (c) Gravitational radius compared to the physical radius for the Earth
((R$_g$/R⊕)) and for the Moon ((R$_g$/R☽)), and ratio of the Earth's
gravitational radius to that of the Moon.

lent objects strike the Earth some 15-20% faster than they do
the Moon.
     It is possible to convert impact velocity to crater dia-
meter D through the energy-scaling relation, which can be
written (Hartmann, 1965)

$$D = CE^k = C(\tfrac{1}{2}MV^2)^k$$

where C and k are constants and the energy E is assumed to be due to the kinetic energy of an object with mass M and impact velocity V. The constant C is not well determined. We use the *ratio* of crater diameters on the Earth to those on the Moon:

$$\frac{D_\oplus}{D_\jupiter} = \left[\frac{(MV^2)_\oplus}{(MV^2)_\jupiter}\right]^k$$

which eliminates the constant C. If objects of equal mass are considered, then the relation becomes

$$D_\oplus = D_\jupiter\left[\frac{V_\oplus}{V_\jupiter}\right]^{2k}$$

where $V_\oplus$ and V are the impact velocities at the surfaces of the Earth and Moon as discussed above.

This relation now depends on k. As discussed by Hartmann (1965), values of k between 1/3.0 and 1/3.4 have been suggested in the literature. Figure 1b is a plot of $D_\oplus$ $D_\jupiter$ as a function of approach velocity for two different values of k. In the velocity range of interest, the resulting diameter ratio is not very sensitive to k, varying only from 1.15 to 1.13 (for k = 1/3.0 to k = 1/3.3, respectively) for the worst case of 15 km/ sec approach. Craters on the Earth will be 11-15% larger than those formed on the Moon by identical objects, as shown in Fig. 1b.

The Earth also collects more of these objects. The lower limit on this is the ratio of the physical cross sections of the Earth and Moon. This goes as the square of the ratio of their physical radii: $(R_\oplus/R_\jupiter)^2 = 3.67^2 = 13.47$. Were there no other considerations, the Earth would gather 13.5 times as many objects, but have the same number of craters *per unit area* as the Moon. But the Earth has a significantly larger *gravitational* radius than the moon, and therefore a larger gravitational cross section. Figure 1c shows the gravitational radius of the Earth and that of the Moon as a function of approach velocity, using a relation given by Wetherill (1974);

$$R_g = R\sqrt{1 + (V_{esc}^2/V^2)}$$

where R_g and R are the gravitational and physical radius of the planet, V_{esc} the escape velocity, and V the approach velocity. Also shown in this figure is the ratio of the gravitational

Table I^a

Basin		Long	Lat	Moon Diam (km)	Area $(10^5 km^2)$	$\%^b$
1	Imbrium	HW–B^c	19	+38	1340	14.100	03.7
2	W Nubium	HW–B	22	–24	425	1.419	00.4
3	Humorum	HW–B	39	–24	410	1.320	00.3
4	Near Schiller	HW–B	45	–56	350	0.962	00.3
5	Bailly	HW–B	68	–67	300	0.707	00.2
6	Grimaldi	HW–B	69	– 5	220	0.380	00.1
7	Pingre	HW–B	78	–56	300	0.707	00.2
8	SE Limb	HW–B	94	–49	480	1.810	00.5
9	Orientale	HW–B	96	–21	620	3.019	00.8
10	Lorentz	HW–B	97	+34	330	0.855	00.2
11	Unnamed B	HW–B	123	+42	410	1.320	00.3
12	Hertzsprung	HW–B	129	+ 1	285	0.638	00.2
13	Birkhoff	HW–B	147	+59	320	0.804	00.2
14	Apollo	HW–B	153	–36	435	1.486	00.4
15	Korolev	HW–B	157	– 4	405	1.288	00.3
16	Unnamed D	HW–B	167	+18	600	2.827	00.7
17	Antoniadi	HW–B	172	–69	140	0.154	00.0
18	Ingenii	HW–B	197	–35	320	0.804	00.2
19	Poincare	HW–B	199	–58	335	0.881	00.2
20	Moscoviense	HW–B	213	+26	410	1.320	00.3
21	Mendeleev	HW–B	219	+ 6	330	0.855	00.2
22	Planck	HW–B	225	–58	325	0.830	00.2
23	Schrodinger	HW–B	228	–75	320	0.804	00.2
24	Unnamed A	HW–B	236	–81	335	0.881	370.3
25	Milne	HW–B	247	–31	240	0.452	00.1
26	Unnamed A	HW–B	252	–68	285	0.638	00.2
27	Compton	HW–B	256	+56	175	0.241	00.1
28	Smythii	HW–B	273	– 1	450	1.590	00.4
29	Humboldtanium	HW–B	278	+58	600	2.827	00.7
30	Crisium	HW–B	301	+16	450	1.590	00.4
31	Fecunditatis	HW–B	312	– 3	520	2.124	00.6
32	Janssen	HW–B	321	–43	160	0.201	00.1
33	Nectaris	HW–B	326	–15	840	5.542	01.5
34	SW Tranquility	HW–B	337	+ 5	350	0.962	00.3
35	Serenitatis	HW–B	342	+25	680	3.632	01.0
A	Sinus Aestuum	SA–B^d	10	+15	250^c	0.491	00.1
B	"Fauth"	SA–B	20	+15	1200^e	11.310	03.0
C	Nubium	SA–B	20	–20	750	4.418	01.2
D	S Procellarum	SA–B	50	–10	710^e	3.959	01.0
E	Question – 1	SA–B	120	+55	f	f	f
F	"Anderson"	SA–B	190	+20	600^e	2.827	00.7
G	"Super Basin"	SA–B	190	–53	2100^e	34.640	09.1
H	"Cyrano"	SA–B	200	–10	480^e	1.810	00.5

Earth (V=17.5 km/sec)			Total		
Diam (km)	Area (10^5km^2)	%^{b}	Area (10^6km^2)	%^{b}	Remarks
1481.3	17.230	00.3	32.100	06.3	
469.8	1.734	00.0	3.229	00.6	inside C
453.2	1.613	00.0	3.005	00.6	
386.9	1.176	00.0	2.190	00.4	
331.6	0.864	00.0	1.609	00.3	
243.2	0.465	00.0	0.865	00.2	
331.6	0.864	00.0	1.609	00.3	
530.6	2.211	00.0	4.119	00.8	
685.4	3.689	00.1	6.872	01.3	
364.8	1.045	00.0	1.947	00.4	
453.2	1.613	00.0	3.005	00.6	
315.1	0.780	00.0	1.452	00.3	
353.7	0.983	00.0	1.831	00.4	
480.9	1.816	00.0	3.383	00.7	inside G
447.7	1.574	00.0	2.932	00.6	
663.3	3.455	00.1	6.435	01.3	
154.8	0.188	00.0	0.350	00.1	inside G
353.7	0.983	00.0	1.831	00.4	inside G
370.3	1.077	00.0	2.006	00.4	inside G
453.2	1.613	00.0	3.005	00.6	
364.8	1.045	00.0	1.947	00.4	inside G
369.3	1.014	00.0	1.888	00.4	inside G
353.7	0.983	00.0	1.831	00.4	inside G
370.3	1.877	00.0	2.006	00.4	inside G
265.3	0.553	00.0	1.030	00.2	
315.1	0.780	00.0	1.452	00.3	inside G
193.5	0.294	00.0	0.548	00.1	
497.5	1.944	00.0	3.620	00.7	
663.3	3.455	00.1	6.345	01.3	
497.5	1.944	00.0	3.620	00.7	
574.8	2.595	00.1	4.834	00.9	
176.9	0.246	00.0	0.458	00.1	
928.6	6.772	00.1	12.610	0215	
386.9	1.176	00.0	2.190	00.4	inside M
751.7	4.438	00.1	8.266	01.1	
276.4	0.600	00.0	1.117	00.2	inside B
1326.5	13.820	00.3	25.740	05.0	overlaps 1
829.1	5.399	00.1	10.060	02.0	
784.9	4.838	00.1	9.011	01.8	
f	f	f	f	f	No basin?
663.3	3.455	00.1	6.435	01.3	Not 16??
2321.5	42.330	00.8	78.830	15.4	
530.6	2.211	00.0	4.119	00.8	

(Continued)

Basin		Long	Lat	Moon Diam (km)	Area $(10^5 km^2)$	%b
I Schwarzschild	SA-B	240	+70	205	0.330	00.1
J Question - 2	SA-B	250	0	f	f	f
K Mare Marginis	SA-B	265	+20	475^e	1.772	00.5
L Mare Australe^i	SA-B	275	-45	900(?)	6.362	01.7
M Tranquility	SA-B	300	+10	750^e	4.418	01.2
N Question - 3	SA-B	350	-20	f	f	f
O Mare Vaporum	SA-B	355	+15	250^e	0.491	00.1
I. Procellarum	IMAB^g	58	+18	900^e	6.362	01.7
II. N Procellarum	IMAB	63	+40	450^e	1.590	00.4
III. "Euclides"	IMAB	28	- 8	450^e	1.590	00.4
a Boltzman	HW-C^h	115	-55	200	0.314	00.1
b Landau	HW-C	119	+42	220	0.380	00.1
c Zeeman	HW-C	134	-75	201	0.317	00.1
d Mach	HW-C	149	+18	205	0.330	00.1
e Gabis	HW-C	152	-14	205	0.330	00.1
f Oppenheimer	HW-C	166	-36	215	0.363	00.1
g Leibnitz	HW-C	182	-38	250	0.491	00.1
h Von Karmann	HW-C	184	-48	240	0.452	00.1
i D'Alembert	HW-C	196	+52	220	0.380	00.1
j Campbell	HW-C	209	+45	235	0.434	00.1
k Gagarin	HW-C	211	-20	270	0.573	00.2
l Fermi	HW-C	237	-19	240	0.452	00.1
m Pasteur	HW-C	255	-12	235	0.434	00.1
aa Clavius	LC	15	-58	230^j	0.416	00.1
bb Sinus Iridium	LC	30	+43	300^j	0.707	00.2
cc Schickard	LC	55	-43	210^j	0.346	00.1
dd Hausen	LC	89	-65	140^j	0.154	00.0
ee Mendel	LC	110	-50	155^j	0.189	00.0
ff Rowland	LC	162	+57	170^j	0.227	00.1
gg Fabry	LC	251	+42	170^j	0.277	00.1
hh Belkovich	LC	272	+62	210^j	0.346	00.1

[a]Total area of Moon covered; 150.200 x 10^5 km^2 or 0.396; total basin area on Earth: 341.870 x 10^6 km^2 or 0.669; all basins and all craters (V_a = 17.5 km/sec). Total area of Moon covered: 119.740 x 10^5 km^2 or 0.315; total basin area on Earth: 272.530 x 10^6 km^2 or 0.533; large, nonoverlapping basins only (V_a = 17.5 km/sec).

[b]Percentage of surface area.

[c]Basin diameters from Hartmann and Wood (1971).

Earth (V=17.5 km/sec)			Total		
Diam (km)	Area $(10^5 km^2)$	%[b]	Area $(10^6 km^2)$	%[b]	Remarks
226.6	0.403	00.0	0.751	00.1	
f	f	f	f	f	No basin?
525.1	2.166	00.0	4.033	00.8	
994.9	7.774	00.2	14.480	02.8	Rim unclear
829.1	5.399	00.1	10.060	02.0	
f	f	f	f	f	No basin?
276.4	0.600	00.0	1.117	00.2	
994.9	7.774	00.2	14.480	02.8	
497.5	1.944	00.0	3.620	00.7	
497.5	1.944	00.0	3.620	00.7	
221.1	0.384	00.0	0.715	00.1	
243.2	0.465	00.0	0.865	00.2	
222.2	0.388	00.0	0.722	00.1	
226.6	0.403	00.0	0.751	00.1	
226.6	0.403	00.0	0.751	00.1	
237.7	0.444	00.0	0.826	00.2	
276.4	0.600	00.0	1.117	00.2	
276.3	0.553	00.0	1.030	00.2	
243.2	0.465	00.0	0.865	00.2	
259.8	0.530	00.0	0.987	00.2	
298.5	0.700	00.0	1.303	00.3	
265.3	0.553	00.0	1.030	00.2	
259.8	0.530	00.0	0.987	00.2	
254.3	0.508	00.0	0.946	00.2	
331.6	0.864	00.0	1.609	00.3	inside 1
232.1	0.423	00.0	0.788	00.2	
154.8	0.188	00.0	0.350	00.1	
171.3	0.231	00.0	0.430	00.1	
187.9	0.277	00.0	0.517	00.1	
187.9	0.277	00.0	0.517	00.1	
232.1	0.423	00.0	0.788	00.2	

[d]Basin diameters from Stuart-Alexander and Howard (1970).

[e]Basin diameters measured by author from Fig. 14 of Howard et al. (1974).

[f]Basins from Stuart-Alexander and Howard (1970), but rim not obvious.

[g]Irregular maria-artificial basin diameter measured by author from Fig. 14 of Howard et al. (1974).

[h]Crater diameters from Hartmann and Wood (1971).

[i]Diameter from Stuart-Alexander and Howard (1970), but rim not obvious.

[j]Large crater diameters measured by author.

[k]Fermi and Tsiolkovsky have combined diameter of 340 km.

radius of the Earth to that of the Moon.   Over the range of
approach velocities of interest, $R_g^{\oplus}/R_g^{☽}$ varies from 4.52 to
4.18, decreasing with increasing velocity.   The Earth's gravitational cross-sectional area is 17.4-20.4 times larger than
the Moon's, compared to the physical cross-section ratio of
13.7.   The Earth therefore collects some 17-20 times as many
objects of a given mass as does the Moon (or 1.3-1.5 times as
many *per unit area*).   If even 30% of the lunar surface was
covered by basins (see below), then at least 45% of the
Earth's surface was disrupted by a similar event 4 billion
years ago.   A more careful estimate is made below.

THE SIZE DISTRIBUTION OF LUNAR BASINS

    The actual number and diameters of lunar basins are not
precisely known.   Table I is compiled from the published lists
of Hartmann and Wood (1971), Stuart-Alexander and Howard (1970),
and Howard *et al.* (1974).   The adopted diameter (column 6) is
generally the "most prominent" ring of Hartmann and Wood, where
available.   Capital letters in column 1 designate basins not
listed by Hartmann and Wood; for some of these, diameters were
estimated from Fig. 14 in Howard *et al.*   Craters larger than
200 km or showing evidence of mare fill are also included in
Table I.   The irregular mare of Oceanus Procellarum is represented by two artificial basins (900 and 450 km across), as is
the irregular mare at 30°W, $+8^{\circ}$ (Euclides).   Mare Gargantua,
which appeared in a recent compilation by Wood and Head (1976),
is not included here; its large diameter (>2500 km) would significantly increase the total area of the Moon covered by basins.   Another very large basin, Super Basin, suggested by
Howard *et al.* (1974), also appeared on the list of Wood and
Head, and is included here.
    Figure 2a is a histogram of the basins and craters from our
list.   The large craters in the 100-199 km diameter range are
underrepresented due to observational selection.   This may
apply to the adjacent bin as well.   Therefore, large *craters*
with diameter less than 300 km were not included in the *minimum*
area count (see below).   In Fig. 2b, we plot a log cumulative
number-log diameter curve for these basins together with the
solid line for highland craters from Hartmann (1966).   Despite
the small numbers, there appears to be a break at D = 500 km,
which is also evident in the histogram (Fig. 2a).   At smaller
diameters, the basins grade into the highland crater curve.
This suggests the lunar surface may be saturated for craters
larger than 500 km; smaller basins and craters may be depop-
ulated by the obliterating effects of one large basin.   Alternative interpretations are also possible; for example, two

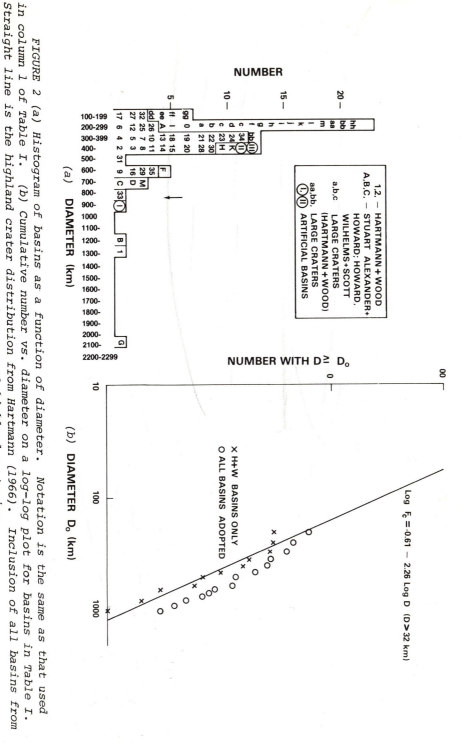

FIGURE 2 (a) Histogram of basins as a function of diameter. Notation is the same as that used in column 1 of Table I. (b) Cumulative number vs. diameter on a log-log plot for basins in Table I. Straight line is the highland crater distribution from Hartmann (1966). Inclusion of all basins from Table I suggests a distribution different from normal highland cratering.

different populations of objects may be represented here. At
the present, there is no way to determine the reality or significance of the 500 km break.

In determining the *minimum* total area of the Moon covered
by basins, it is necessary to eliminate overlap between basins.
Where small basins lie *inside* larger areas (see remarks column
in Table I), the small basins were discarded entirely (for
example, basins 14, 17, 18, 19, 22, 23, 24, and 26 were not
counted for this reason). When two basins overlap (for example,
Fauth = basin B overlaps Imbrium = basin 1), the effective
diameter of the smaller basin was decreased so as to count only
the nonoverlapping *area* of each basin. Mare Australe, whose
rim is difficult to identify and whose diameter is therefore
uncertain, was also eliminated in the miniumum basin area count.

The total area of the Moon covered by large, nonoverlapping
basins is 32% of the available surface area. Inclusion of *all*
basins and large craters in Table I raises this figure to 40%.

BASIN FORMATION ON THE EARTH'S SURFACE

The Earth should have collected 1.3-1.5 times as many objects per unit area as did the Moon. Each basin on the Earth
was 11-15% larger than the corresponding crater on the Moon
(for objects of the same mass). This means each terrestrial
basin had roughly 28% more *area* than would the lunar basin
produced by an equivalent object. Figure 3 shows the total
area of the Earth's surface covered by basins as a function of
approach velocity for all basins and for nonoverlapping large
basins only. For 32% of the lunar surface covered by basins,
the corresponding figure for the Earth is 48-62% of the surface
area. If we adopt the larger 40% coverage of the Moon, 60-78%
of the Earth's surface was affected by basin formation. Column
10 of Table I gives the area each lunar basin would have on the
Earth for an approach velocity of 17.5 km/sec. Summation of
this column yields 67% coverage of the Earth if *all* basins are
counted, 53% if only nonoverlapping, large basins are included.

Therefore, 4 billion years ago, *at least* 50% of the surface
of the Earth was disrupted by basin-forming impacts. This
figure represents the *minimum* percentage of a global crust
affected. No attempt has been made to account for the overlapping of terrestrial basins (which must have occurred); but
the above figure is based on an absolute minimum number of nonoverlapping basins *observed* on the Moon. The actual number of
basin-forming objects impacting the Moon was certainly much
higher than the number of surviving basins. For example, if
Super Basin had occurred *last* on the Moon it would have destroyed at *least* eight basins with diameters between 280 and

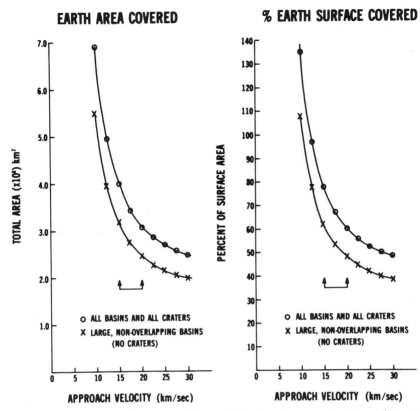

FIGURE 3. *Total area affected by basins as a function of approach velocity for the Earth. Two cases are included: a minimum number of large, nonoverlapping basins, and all basins and craters from Table I. (b) Percentage of Earth's surface disrupted by basin-forming impacts. For low velocities, the curves exceed 100% because there has been no correction for overlap.*

435 km. There is no way to correctly estimate the number of smaller basins eradicated by the formation of an Imbrium or Orientale.

Another potential problem we have not investigated is the possible shadowing and/or focusing effect the Moon may have had on Earth-bound objects. If the flux duration was long compared with the orbital period of the Moon 4 billion years ago, the effect is probably small unless the Moon was very close to the Earth.

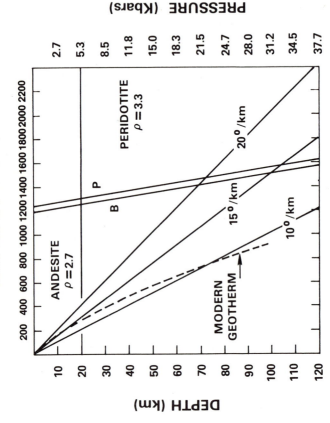

FIGURE 5. A pressure-temperature model for the Earth 4 billion years ago, as suggested by Frey and Lowman (1976). The crust is andesite and 20 km thick. The mantle is solid peridotite down to ∼105 km. The adiabatic pressure-melting curves for peridotite (P) and basalt (B) are shown. Several geothermal gradients are also plotted.

EFFECTS OF BASIN-FORMING IMPACTS

The impact of a large object produces two immediate effects:
the excavation and ejection of large amounts of crustal material, and the fracturing and brecciation of crustal rocks to
great depths. The former is the more obvious, but the latter
affects a greater volume of rock. Baldwin's (1963) data suggest a 500 km basin will have a depth of ∿9.5 km. This is
based on extrapolation of smaller craters and *observed* diame-
ter-depth ratios, and therefore represents the *minimum* depth
of the original crater. Basin depth modification results from
back-falling ejecta, slumping and mass-wasting of the walls,
isostatic adjustment of the basin topography, and possible
filling by mare-type liquids.

Fragmentation of the underlying rock is severe. Data from
Table XII of Baldwin (1963) and from Innes (1961) suggest a
relation between the depth to the bottom of the brecciated
layer ($\underline{B}$) and the crater diameter ($\underline{D}$):

$$\log \underline{B} = 1.0232 \log \underline{D} - 0.5905$$

which is a least-squares fit to the above data. This is shown
in Fig. 4. The inset gives the results for *terrestrial* craters
where drill cores have provided direct measurements. The curve
is then extrapolated into the diameter range for basins. The
two orders of magnitude extrapolation are probably not valid,
but are the only available information at present. A 500 km
basin impact excavates almost 10 km of crust but fractures rock
to depths of ∿150 km.

To understand the effects of such an impact, a model of the
crust and mantle for the Earth 4 billion years ago is needed.
We adopt the following (Frey and Lowman, 1976): The crust is
andesite with a bulk density of 2.7 gm/cm^3. The crust-mantle
boundary is 20 km deep, which is consistent with Condie's
(1973) suggestion that the Archaean crust thickened to 25-30 km
between 3.5 and 3.0 billion years ago. The mantle is solid
periotite with a density of 3.3 gm/cm^3. These relations are
shown in Fig. 5. Geothermal gradients were probably higher in
the past. We adopt 15°K/km, compared with the present day
value (shown dashed) of roughly 10°K/km. The pressure melting
curves for basalt (the partial melt product) and peridotite
are also shown. The latter intersects the thermal gradient
curve at roughly 105 km; below this the mantle is molten in
this simple model.

Basin impacts initially produce a dichotomy in elevation;
basin floors are ∿10 km below the original crust for 500 km impacts. We might expect two subsequent effects: isostatic adjustment of the basin topography, which tends to smooth out

the elevation differences, and basaltic filling of the basin
(see below), which would produce a compositional dichotomy between basin floor and highland crust. The relative time
scales are important.

Below the basin the pressure-temperature relations are
changed in the sense that melting is favored at shallower
depths. Material rising upward in response to isostatic pressures finds itself in a region where its temperature exceeds
the pressure-melting temperature for that level. Partial melting produces a basaltic liquid with a density of $\lesssim 3.0$ compared
to the solid but fragmented 3.3 peridotite above. The liquid
will clearly rise in this situation, as depicted in Fig. 6.

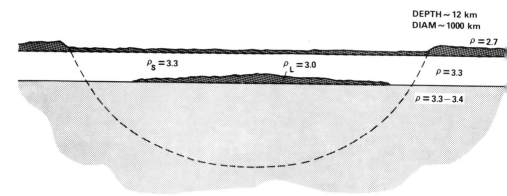

FIGURE 6. *Mare-type flooding of a large impact basin will
be rapid on the Earth. Formation of the basin changes the
pressure-temperature relations below the basin, and favors melting at shallower depths. The resulting liquid basalt is less
dense ($\sim$3 gm/cm³) than the overlying fractured rock (3.3 gm/cm³).
The rise time is short compared to the isostatic adjustment of
the basin topography (see text).*

The rising material promotes further melting. Near the surface
hydrostatic pressure expands the liquid into the crater, flooding the basin. The rise time for the lava, based on seismic
studies of Hawaiian basaltic eruptions, is on the order of 10-
100 years. This time is short compared with isostatic adjustment times (several *thousand* years) based on Fennoscandian uplift from postglacial rebound. While the thinner crust of the
early Earth would tend to shorten this period, the adjustment
time will still be long compared with the time scale for
flooding. The full isostatic adjustment is $\sim$80% of the original
depth of the crater, producing a 2-3 km height difference between the cratered highlands and the basalt-flooded floor of
the basin. If allowance is made for basin-ejecta thicknesses,
the final elevation difference is 3-4 km.

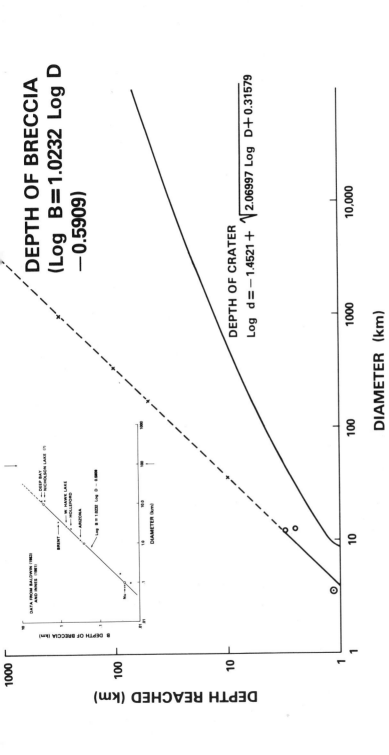

FIGURE 4. Depth reached by excavation (depth of crater) and by fragmentation (depth of breccia) as a function of basin diameter. Crater depth is based on Baldwin's (1963) formulas. Depth to bottom of breccia is an extrapolation of the least-squares fit shown as the inset, which is based on terrestrial crater data as reported in Baldwin (1963) and Innes (1961). The extrapolation is shown as the dashed line.

There is evidence of extensive volcanism early in the history of the Earth. The ancient greenstone belts contain members produced by 60-80% melting of peridotitic material, such as might occur due to the pressure drop caused by large impacts (Green, 1975). Glickson (1976) points out that these volcanic rocks must postdate the basin-forming period since the rocks contain no shock effects, and Green (1972) argues that these extensive eruptions are similiar to the lunar maria.

DISCUSSION

The period of basin formation was therefore capable of converting more than 50% of an original global low-density crust into a mare-highland type of dichotomy. With subductable crust now present, plate tectonic processes could begin and probably did so within a short time (Frey, 1977a). These processes have subsequently reworked, over many cycles, the maria of the Earth. This reworking of the high-density part of the Earth's crust produced changes for the remnant highland patches. These blocks were carried about by seafloor spreading, colliding, and welding during the Archaean, eventually stabilizing into large units that became the continental shields. High temperatures during these collisions probably account for the relatively young age of the surviving rocks. For the last 2.5 billion years or so, large-scale modern plate tectonics (large plates, moderate spreading rates) has replaced the earlier fast spreading of microplates (Frey, 1977a; Burke et al., 1976). But for the last 4 billion years, according to the above analysis, there have been types of ocean basins with depths comparable to those of today. If such basins were important in the origin of life on the Earth, then it becomes important to ask whether such basins are expected on other terrestrial planets in other star systems.

Chapman (1976) and Murray et al. (1975) argue that the lunar and mercurian basins were formed from objects that were common to the entire inner solar system about 4 billion years ago. If so, then among the most plausible orbits for these objects are ones that require them to originate in the outer part of the solar system. This storage in the Uranus-Neptune region allows the 500 million year delay between the formation of the terrestrial planets and their bombardment by this flux of asteroidal bodies. These objects were perturbed sunward, eventually crossing the orbital space of the terrestrial planets. Thus the basin-forming events were dependent not only on the conditions of formation of the planets and planetesimals, but also on the particular distribution of planetary masses in our system. Therefore, not only the event itself, but also

its timing, depend on the size and location of the other planets, particularly Jupiter. There is no reason therefore to expect such a period of massive crustal destruction by asteroidal
bodies on otherwise terrestrial planets in other planetary
systems where the arrangement of the planetary masses could be
quite different.

Two extreme cases can be considered: (a) the basin-forming
event does not occur at all in another planetary system, or (b)
it does occur but at a time very different (in terms of planetary history) from the time in our system.

If the event does *not* occur on planets with an initially
global low-density crust, then these planets retain a relatively
intact continental mass completely encircling the planet. Geologic evolution would be significantly different on such a
world. Plate tectonic processes should not occur on such a
world unless subductable crust becomes available by some other
means. The crust may become broken into large or small plates,
but these should not move in horizontal directions over any significant distance. The plates might be expected to jostle
about with minor compression occuring between slowly colliding
blocks. Major seafloor spreading would not exist, and the huge
folded mountain belts of the Earth would be absent. Volcanics
would dominate orogeny. These effects have been discussed
elsewhere (Frey, 1976b). Here we are more concerned about the
effect such a different physical environment (no deep basins)
would have on the development of life. This will depend in
part on the volatile budget of the planet and on its surface
temperature. If liquid water is abundant, a planet with a
global crust should also have a global ocean covering that
crust. A water-poor world would be dry with a few patches of
water perhaps forming in the larger craters on the surface.
Neither of these situations precludes the origin of life, but
both environments suggest very different evolutionary patterns
which may have little in common with the evolution of life on
the Earth.

In particular, the global ocean world would have no evolution of land-dwelling forms, and even within the ocean there
may be a lack of variety of lifeforms because of the relative
stability of the physical conditions there. The water-poor
world is a restrictive environment as well. If temperature
extremes occur, the smallish patches of water may periodically
dry out and reform, seasonally perhaps, or freeze and thaw on
a regular basis. The climate of such a world will certainly be
different than that on the Earth because of the absence of the
heat sinks provided by the large, deep oceans here. In both
cases of planets on which basins did not form there is good
reason to believe that evolutionary patterns will depart significantly from that of the Earth; life may well be present
but it might very well be very different from that of our own
world.

A different situation exists for those worlds on which the basin-forming events do happen but at very different times than on the Earth. In our case the basin-forming bombardment occurred after the crust was thick enough to record the event; that is, the basin topography was preserved as a significant elevation difference. The event also occurred early enough in the history of the Earth so that the long sequence of events leading to the formation of life could begin at an early stage, finally producing complex, advanced lifeforms in less than 4 billion years of planetary history. If the event occurs so early on another planet that the crust has not thickened to the point of preserving the crater, then the effects of the impact are lost as the surface anneals itself after the large objects punch through to shallow molten layers; it is as if the event never happened, and this extreme of early basin-formation reduces to the cases of intact global crusts discussed above. The more distressing case is that of *late* basin formation on a world where life has become established. If a major series of catastrophic impacts occurs at the rate of one every million years (on the Earth, several hundred major basins in a few hundred million years), this could lead to significant changes in the pattern of evolution. The long duration between, and random location of, the impacts do not suggest that life would be obliterated on such a world, except locally. But such short-term disruptions of the crust and environment could well destroy newly developing organisms, thus setting back the development of more complicated and perhaps more intelligent life-forms. Life is very sensitive to even small changes in the physical environment and a basin-forming impact, although a brief event, is not a small change near the point of impact. The pattern of evolution of such a world could well have been quite different from that of the Earth, and the possibility of very different physical environments on other terrestrial planets should be carefully considered in discussions of the origin of life that use the Earth as an example. Ocean basins may well be rare on terrestrial planets.

CONCLUSIONS

The origin of the Earth's ocean basins may be traced to a period of intense cratering by basin-forming objects some 4 billion years ago. These impacts are capable of converting more than 50% of an original global low-density crust (a common characteristic of terrestrial planets?) into a mare-highland crustal dichotomy (high-density/low-density crust). This dichotomy has been preserved as plate tectonic processes have reworked the maria into modern oceanic crust. Deep basins have existed on the Earth for most of its history.

Because the events suggested here depend on the exact configuration of planetary masses, not only for the impacting objects but also for the timing of the impacts, there is no reason to believe that such events are common on terrestrial planets in other star systems. Global crusts may be common, but ocean basins of the sort found on the Earth may be rare. In the case of no basin formation or of a very late timing of such an event, the physical environment in which life originates and evolves is quite different from that of the Earth. Although lifeforms probably arise in each of the environments described above, the evolutionary pattern of these lifeforms is likely to be very different from that of the Earth. Physically diverse environments on otherwise similiar planets should be considered carefully when the Earth is used as a standard in discussions of the origin and evolution of life, for there is no guarantee that the most basic attribute of the Earth - its deep but not globe-encircling oceans - is common in the universe.

ACHNOWLEDGMENTS

It is a pleasure to thank Paul Lowman for suggestions and encouragement during tihis work and for fruitful collaboration on the evolutionary style of the Earth, and to thank Michael A'Hearn and Ernst Öpik for thoughtful reviews of an earlier version on this manuscript.

REFERENCES

Baldwin, R. B. (1963) "*The Measure of the Moon.*" Univ. of Chicago, Illinois.
Burke, K., Dewey, J. F., and Kidd, W.S.F. (1976). Dominance of horizontal movements, arc and microcontinental collisions during the later permobile regime, in, "*The Early History of the Earth*" (B, ,. F. Windley, ed).
Chapman, C. R. (1976). Chronology of terrestrial planet evolution: The evidence from Mercury, *Icarus 28*, 523-536.
Condie, K. C. (1973). Archean magmatism and crustal thickening, *Geol. Soc. Am. Bull. 84*, 2981-2992.
Engel, A. E. J., Itson, S. P., Engel, C. G., Stickney, D. M., and Cragy, E. J. (1974). Crustal evolution and global tectonics: A petrogenic view, *Geol. Soc. Am. Bull. 85*, 843-858.
Frey, H. (1977a). Ph. D. Thesis, Univ. of Maryland.
Frey, H. (1977b). Origin of the Earth's ocean basins, *Icarus 32*, 235-250.

Frey, H. and Lowman, P. D. (1976). Impact Basin formation and the early terrestrial crust, (Abstract) of paper presented at Division for Planetary Sciences Meeting in Austin, Texas, April 1976.

Glickson, A. Y. (1976). Earliest Precambrian Ultramafic-mafic volcanic rocks: Ancient oceanic crust or relic terrestrial maria? *Geology 4,* 201-205.

Green D. H. (1972). Archaean greenstone belts may include terrestrial equivalents of lunar maria: *Earth Planet. Sci. Lett. 15,* 263-270.

Green, D. H. (1975). Genesis of archaean peridotitic magmas and censtraints on archaean geothermal gradients and tectonics, *Geology 3,* 15-18.

Hartmann, W. K. (1965). Terrestrial and lunar flux of large meteorites in the last two billion years, *Icarus 4,* 157-165.

Hartmann, W. K. (1966). Early lunar cratering, *Icarus 5,* 406-418.

Hartmann, W. K. and Wood, C. A. (1971). Moon: Origin and evolution of multiringed basins, *The Moon 3,* 3-78.

Howard, K. A., Wilhelms, D. E., and Scott, D. H. (1974). Lunar basin formation and highland stratigraphy, *Rev. Geophys. Space Phys. 12,* 309-327.

Innes, M. J. S. (1961). The use of gravity methods to study the undergraound structure and impact energy of meteorite craters, *J. Geophys. Res. 66,* 2225-2239.

Lowman, P. D. (1976). Crustal evolution in silicate planets: Implications for the origin of continents, *J. Geol. 84,* 1-26.

McKenzie, D. P. (1973). Speculations on the consequences and causes of plate motions, *in,* "Plate Tectonics and Geomagnetic Reversals," pp. 447-469. Freeman, SanFrancisco.

Murray, B. C., Strom, R. G., Trask, N. J., and Gault, D. E. (1975). Surface history of Mercury: Implications for the terrestrial planets, *J. Geophys. Res. 80,* 2508-2514.

Opik, E. J. (1966). The stray bodies in the solar system.II, *Adv. Astron. Astrophys. 4,* 301-336.

Stuart-Alexander, D., and Howard, K. (1970). Lunar maria and circular basins - A review, *Icarus 12,* 440-456.

Tera, F., Papanastassiou, D. A., and Wasserburg, G. J. (1974). Isotopic evidence for a terminal lunar cataclysm, *Earth Planet. Sci. Lett. 22,* 1-21.

Wetherill, G. W. (1974). Solar system sources of meteorites and large meteroids, *Ann. Rev. Earth Planet. Sci. 3,* 303-331.

Wetherill, G. W. (1975). Pre-mare cratering and early solar system history, *Proc. Soviet-American Con. Cosmochem. Moon Planets.*

Wetherill, G. W. (1976). Comments on the paper by C. R. Chapman: Chronology of terrestrial planet evolution The evidence from Mercury, *Icarus 28,* 537-542.

Wise, D. U. (1974). Continental margins, freeboard and the volumes of continents and oceans through time, *in,* The Geology of Continental Margins" (C. A. Burk and C. L. Drake eds.), pp. 45-58. Springer-Verlag, New Park, New York.

6

A STRATIGRAPHIC APPROACH
TO THE EVOLUTION OF THE LUNAR CRUST[1]

Farouk El-Baz

National Air and Space Museum
Smithsonian Institution, Washington, D.C.

The original lunar crust was physically and chemically modified by meteorite bombardment and the subsequent transport and
redistribution of its materials. The exposed relics of this
crust form the lunar highlands, which are predominantly sculptured by over 30 large basins that range in diameter between
300 and 2000 km. These basins are equally distributed on the
near and far lunar hemispheres. The largest, the South Pole/
Aitken basin on the far side, is among the oldest and is recognized by a subdued, discontinuous mountain chain surrounding a
depression about 6 km deep. The youngest, the Orientale basin
on the western limb, displays distinct mountain rings, a continuous lineated ejecta blanket surrounding the outer ring (930
km), and smooth plains and secondary craters farther out. Lunar basins display a continuous array of morphologies, from extremely subdued to fairly crisp features. This negates the possibility of sculpting the crust by meteorite impacts in one
catastrophic event. Lunar crustal materials most likely were
bombarded continuously by large objects throughout the early
lunar history, or between 4.6 and about 3.9 billion years ago.
In addition to impact products, some highland units may be volcanic in origin, including light plains on the lunar far side
and spectrally distinct domes on the near side.

Although stratigraphically very significant, mare materials
cover less than 20% of the lunar surface and compose only 1% of
the crust. Mare basalts were deposited within the lunar basins
and surrounding troughs during three or four major episodes,

[1]This work was performed under NASA grant NSG-7188.

103

between 3.85 and about 2.5 billion years ago. Basaltic vol-
canism indicates not only the chemistry of the crust, but also
the amount and length of time of internal planetary differen-
tiation.

INTRODUCTION

Stratigraphy is the study of the formation, composition,
and correlation of surface rocks in space and time. Application of the principles of stratigraphy unravels the sequence
of events that constitute the geologic history of a region or
a planetary body. These principles grew from observation of
terrestrial processes, but many are applicable to the Moon as
well.

Two laws are basic to both lunar and terrestrial stratigraphy: (1) the law of superposition: when one rock unit is
superposed on another, the uppermost is younger than the one
it overlies. This law holds true if the rock sequence has not
been disturbed or overturned since its formation. (2) The law
of cross-cutting relationships: a geologic feature is younger
than the unit it cuts across.

The Moon displays other features, unavailable on Earth, for
deciphering stratigraphic relationships. One of these is the
nature and frequency of craters on a given surface area. The
older the surface, the longer it has been exposed to meteorite
bombardment. Therefore, an older surface will generally display
a greater number of craters, and/or a degraded and subdued
appearance. Relative ages of different mare units can be established by analysis of crater diameter/frequency relations.
In crater analysis, however, care must always be exercised to
avoid confusing primary craters with secondary impact or volcanic craters.

Distinction between impact and volcanic craters on the Moon
has long been a matter of controversy. Earth-based photographs
provided the basis for the opposing theories of lunar crater
formation by meteorite impact (e.g., Gilbert, 1893, 1896; Shoemaker, 1962; Baldwin, 1963) and by volcanic processes (e.g.,
Dana, 1846; Spurr, 1944, 1945; Green, 1962). Although the controversy is still with us, it is now feasible to photogeologi-
cally distinguish between features that are the product of meteorite impacts and those created through internal lunar processes. Close-up Lunar Orbiter photographs and those taken
later by sophisticated cameras aboard Apollo missions 15 through
17 are used to distinguish characteristics of products of both
processes (McCauley, 1968; Mutch, 1972; El-Baz, 1974). The
fact that most lunar craters were formed by meteorite impacts
was supported by studies of terrestrial meteorite impact craters

(Shoemaker, 1962; Fielder, 1965), underground explosions (Short, 1965; and Dence, 1972), and laboratory simulations (Gault et al., 1968; and Gault, 1970). The impact theory was confirmed by results of the Apollo lunar missions (see reviews by El-Baz, 1975; Taylor, 1975; Short, 1975; King, 1976; and Lindsay, 1976).

Another stratigraphic tool that dominates the lunar panorama, but is lacking on Earth, is the presence of large multiringed basins of impact origin. Because these basins and their deposits cover large areas, they provide a widespread reference for correlating relative ages of surface units. Basins typically exhibit concentric rings of prebasin material disturbed by the impact, a hummocky ejecta blanket that thins with distance from the basin, and smooth plains and secondary craters beyond that blanket. The number of rings present depends on the basin size. Not all basins have mare fill; farside basins have little or none. There are over 30 basins on the Moon ranging in size between 300 and 2000 km in diameter. They vary in appearance from sharp and fresh to degraded and barely discernible.

On the basis of these morphologic characteristics the basins have been ranked according to relative age (Stuart-Alexander and Howard, 1970; Hartmann and Wood, 1971). The superposition and cross-cutting relationships of basins to other lunar features and units can therefore help define a lunar wide sequence of events. For example, it is clearly recognized that the Orientale basin is younger than the Imbrium basin, and that the Imbrium basin is in turn younger than the Nectaris basin.

THE LUNAR STRATIGRAPHIC COLUMN

Most of the Earth-facing lunar hemisphere was geologically mapped by the U.S. Geological Survey during the 1960s using telescopic photographs. This mapping was based on the premise that the lunar crust displayed material units that could be defined by physical properties and ranked in order of decreasing age.

Although lunar stratigraphy is still a relatively new field, several attempts have been made to classify features by relative age according to topographic and morphologic characteristics. The basic lunar stratigraphic sequence was first developed by Showmaker and Hackman (1962) in an area that includes the southern part of the Imbrium basin. This sequence was later modified and applied to most of the near side (McCauley, 1967; Wilhelms, 1970; Wilhelms and McCauley, 1971; Mutch, 1972).

An example of deducing lunar surface stratigraphy is in
Fig. 1. Depicted in this sterogram are remnants of the rims
of the craters Fra Mauro, Bonpland, and Parry. In the upper
left corner of the area there is a thick deposit with radial
segments, which is believed to be ejecta from the Imbrium basin to the north of the area of the sterogram. This deposit is
superposed on the high rim of the crater Fra Mauro and therefore is younger than that rim. This lineated deposit has been
named the Fra Mauro Formation (Eggleton, 1964). Crater interiors, as well as a region to the lower right, are occupied by
a relatively flat, light-colored unit. This unit has also been
interpreted as ejecta from the Imbrium basin and has been named
the Cayley Formation (Wilhelms, 1965). It displays a well-developed system of straight rilles. Since the Parry Rilles cut
across both the crater rims and the light-colored fill, these
grabens are younger than both. A crater in the center of the
photograph to the right obliterates a rille, and its ejecta
overlies the light-colored (Cayley) unit. This relationship indicates that material of that particular crater is younger than
the episode that resulted in the deposition of the Cayley unit
as well as the tectonic movements that caused the rilles. The
lower left part of the area is occupied by a dark, relatively
smooth unit that displays fewer small craters than on the Cayley
fill. This dark mare material shows flow fronts and wrinkly
ridges. It overlaps and embays the Cayley unit and truncates
and covers one rille in the center of the photograph on the
left; it is therefore younger than both.

Hence the following sequence of events may be constructed.
First to have formed are the three large craters, probably in
the following sequence: Fra Mauro, Bonpland, and Parry. These
craters probably predate the formation of the Imbrium basin to
the north, because they are covered by the lineated unit that
is believed to be ejecta from Imbrium (the Fra Mauro Formation).
The craters were filled by finer ejecta from Imbrium and other
impact events as well as locally generated debris, which formed
the light-colored fill (Cayley Formation). This filling started
prior to the formation of the Imbrium basin was nearly completed
during that event, but continued after it. Tectonic movements
caused the formation of Parry Rilles by the collapse of rock between two parallel faults. At a later time, lava flooded
troughs southwest of the region and the flows covered the low-
land, including terminal portions of some rilles. Finally, after the emplacement of the mare material, impact events created
numerous craters in the region; some of these, showing very
crisp features, are younger than others.

FIGURE 1. Stereopair showing remnants of the rim of the crater Fra Mauro (upper edge) and below it the craters Bonpland (left) and Parry (right). The lineated unit in the upper left corner is part of the Fra Mauro Formation, an ejecta deposit radial to the Imbrium basin. The crater rims and their light-colored fill are crossed by Parry Rilles, a system of forked linear depressions. The dark mare unit in the lower left area is superposed on, and therefore younger than, the three craters, the Fra Mauro Formation, the light plains, and the graben rilles. (Apollo 16 metric photographs 1980 and 1981.)

FIGURE 2. Three faces of the Moon showing the major stages of evolution of the near side: first, the formation of the Imbrium basin whose materials are superposed on earlier features; second, the flooding of the basins and nearby lowlands by mare materials; and third, the formation of postmare craters (after Wilhelms and Davis, 1971).

Sequences such as this can be worked out for many parts of the lunar surface. They are particularly clear where mare units abound. Within the older lunar highlands, however, there are many more complications that cloud regional as well as local stratigraphic sequences. However, it is possible to generalize the stratigraphic sequences on the near side of the Moon based on the Imbrium basin formation (Fig. 2). Also, it was recently realized that the nearside lunar stratigraphy could be extended, with modifications, to the far side as well (Stuart-Alexander and Wilhelms, 1974; El-Baz, 1975; El-Baz and Wilhelms, 1975).

The Moon-wide, time-stratigraphic sequence in order of decreasing relative ages is as follows:

1. *Pre-Nectarian* All materials formed before the Nectaris basin and as long ago as the formation of the Moon are classed as pre-Nectarian. Most pre-Nectarian units are present on the lunar far side. These include materials of very old and subdued basins, and mantled and subdued craters.

2. *Nectarian System* This system includes all materials stratigraphically above and including Nectaris basin materials, and up to but not including Imbrium basin strata. However, in much of the area surrounding the Imbrium basin, Nectarian basin materials cannot be recognized and the pre-Nectarian and Nectarian can be combined as pre-Imbrian materials. Ejecta of the Nectaris basin that can be traced near the east limb region allow recognition of these materials as an important stratigraphic datum for the farside highlands. Some light-colored plains units, particularly on the far side, are believed to be Nectarian in age.

3. *Imbrian System* A large part of the lunar surface is occupied by ejecta surrounding both the Imbrium and Orientale basins. These form the lower and middle parts of the Imbrian System, respectively. They include the Fra Mauro Formation and several patches of light-colored plains. Two-thirds of the mare materials belong to the Imbrian System, particularly in the eastern maria of Crisium, Fecunditatis, Tranquillitatis, Nectaris, and the dark annulus of Serenitatis, as well as most mare occurrences on the lunar far side.

4. *Eratosthenian System* This system includes materials of rayless craters such as Eratosthenes. Most of these are believed to have once displayed rays that are no longer visible because of mixing due to prolonged micrometeoroid bombardment and solar irradiation. The system also includes about one-third of the mare materials on the lunar near side. These are generally concentrated in Oceanus Procellarum, in western Mare Imbrium, and possibly in the central region of Mare Serenitatis.

5. *Copernican System* This is stratigraphically the highest and, hence, the youngest lunar time-scale unit. It includes materials of fresh-appearing, intermediate to high albedo,

bright-rayed craters. The system also includes exposures of
very high albedo material on the inner walls of craters and
scarps. Brightness in these cases is believed to result from
fresh exposure by mass wasting and downslope movement along
relatively steep slopes. The Copernican System also includes
isolated occurrences of relatively small, dark-halo craters.
Although some of these are probably impact craters, others may
be volcanic in origin.

THE HIGHLANDS

     When Galileo Galilei observed the Moon through a telescope
for the first time, he noticed a dichotomy between highlands
and maria. This dichotomy was confirmed by both photogeologic
interpretations and the study of returned lunar samples.
     Highland materials cover nearly 83% of the lunar surface,
including most of the lunar far side. They are composed mainly
of light-colored, low-density, feldspar-rich impact breccias.
     Basins and large craters dominate the highland physiography.
In many cases overlap relationships between basins and large
craters are clear, and in some cases superposition relationships
are not easy to decipher. As an example of decreasing numbers
of craters with age, in the central far side region of the
Moon (between 50° north and south and 140° east and west),
Stuart-Alexander (1976, p. 9) recognized the following sequence
of craters 100 km or more across: 47 pre-Nectarian craters, 38
Nectarian age craters, and 14 Imbrian age craters.
     Large, multiringed lunar basins vary greatly in the degree
of preservation. Many are indicated only by a few subdued and
discontinuous mountain rings. The Orientale basin on the west
limb of the Moon (Fig. 3) is the youngest of lunar basins. Its
fresh appearance and its sparse mare fill make it the best preserved example of lunar basins. Knowledge gained from study
of Orientale can therefore be extrapolated to the other less
distinct structures.

The Orientale Basin

     Photographs of the Orientale basin, which is centered at
$96^{\circ}W$, $21^{\circ}S$, have been most convincing in deducing theories of
origin of lunar basins. Although the first photographs were
provided by Zond 3, the Lunar Orbiter IV images have revealed
the critical information that this basin is probably the youngest of all the large lunar depressions (Fig. 3). Its features
are crisp and sharp, and resulted in numerous interpretations.

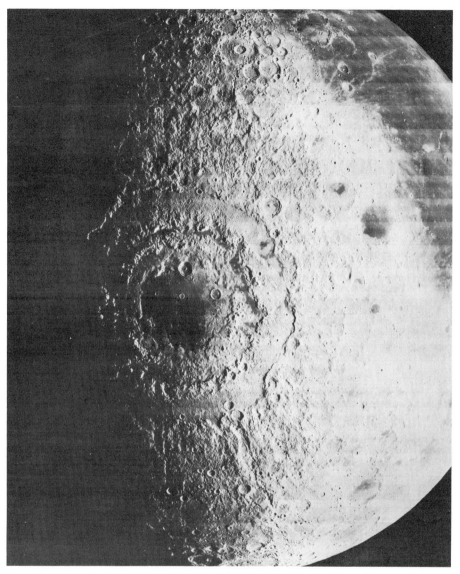

FIGURE 3. The multiple rings of the Orientale basin. Four major circular rings are displayed; the innermost is 320 km in diameter and the outermost is 930 km in diameter. Note the radial ejecta blanket surrounding the outer ring. (Lunar Orbiter IV frame M-187.)

The basin displays four well-defined rings and a possible fifth ring (Hartmann and Wood, 1971). The four main rings are as follows:

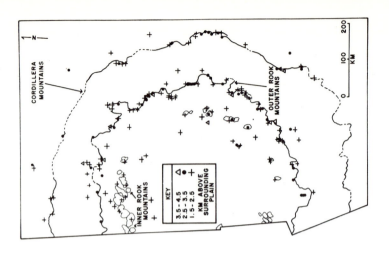

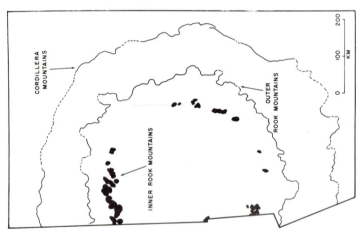

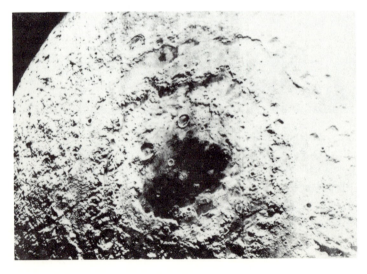

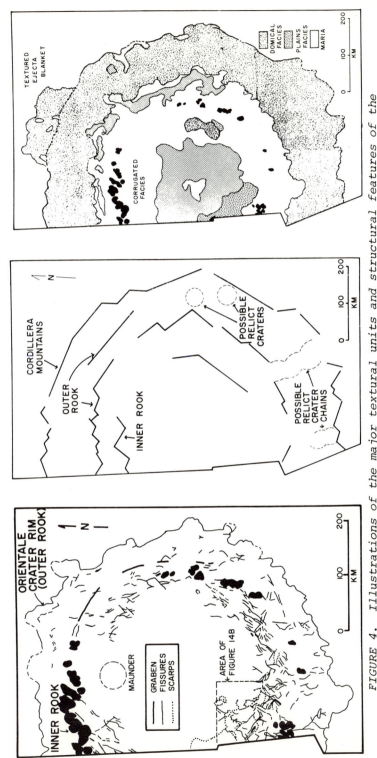

FIGURE 4. Illustrations of the major textural units and structural features of the Orientale basin (after Head, 1976a).

(1) An inner ring, 320 km in diameter, which encloses most of the dark, basaltic material of Mare Orientale. This ring is here called the "Maunder Ring" after a well-defined, super-posed crater on the nothern part.

(2) The second ring, 480 km in diameter, which is marked by the inner Montes Rook, and is surrounded on the east side by mare materials of Lacus Veris.

(3) The third ring, 620 km in diameter, which is marked by outer Montes Rook.

(4) An outer ring, 930 km in diameter, which is bounded by Montes Cordillera.

The possible fifth ring (up to 1300 km in diameter) is only suggested by a few discontinuous and barely visible scarps.

Interpretation of basin stratigraphy and thickness of ejecta depends upon the mode of formation of basin rings. Al-though textural and structural characteristics of Orientale are easy to decipher (Fig. 4), there is no agreement as to how the rings formed, and consequently, how much ejecta was excavated by the impact event. Also, since Orientale is the best pre-served example of the lunar basins, its deposits have been sub-ject to many interpretations, resulting in an abundance of nomenclature (Table I).

The concentric rings of the Orientale basin are related one way or another to the impact of a large body at the center of the basin. The earliest interpretations of these rings ascribed them to collapse along concentric faults (Hartmann and Kuiper, 1962). McCauley (1968) proposed that during the early stages of shock wave propagation that followed the instant of impact there was compressive failure and uplift of large structural blocks. Some uplifts may have occurred along concentric in-ward-dipping thrust faults. The rings may therefore represent frozen shock waves, or "frozen tsunami-like waves" (Baldwin, 1972). Alternatively, these immense mountain chains may have resulted from later readjustment of the terrain by upward thrust along the circular faults created by the impact event.

Most investigators agree that the mare material within the Maunder ring was emplaced well after the basin event, and that the more hummocky and fractured material within the Rook Moun-tains was formed as a shock melt (Moore et al., 1974) during the impact event. Similarly, most investigators agree that the basin deposits outside the Cordillera ring were formed as ejecta during the basin excavation. Well-developed radial lin-eations and alignment of secondary craters support this inter-pretation, and the outer fringes of Orientale may be subdivided into several textural units (Moore et al., 1974; McCauley, 1976).

The interpretation of the region between Montes Rook and Cordillera, however, is not as simple (Fig. 5). The terrain

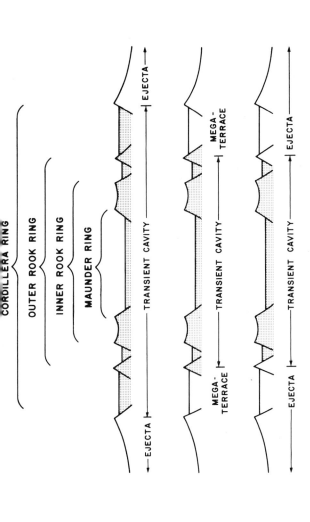

FIGURE 5. Schematic illustrations of three theories of origin of the Orientale basin rings. Top: according to Hodges and Wilhelms (1976a, b), the transient cavity is marked by the Cordillera ring where the multiple ring structure may be caused by uplift of discrete resistant layers; middle: according to Head (1976a) the outer Rook ring constitutes the crater rim where a mega-terrace is assumed to form the Cordillera ring; bottom: according to McCauley (1976), the outer Rook mountains mark the transient cavity, since lineated ejecta can be traced inward to their rims.

TABLE I. Correlation of Nomenclature of the Orientale Basin and Related Materials.

Center of basin	Maunder ring	Inner rook ring	Outer rook ring	Cordillera ring	Source
radius in km	→160	240	310	465	
Central basin plains and mare material			Montes rook formation	Cordillera formation	McCauley (Mutch, 1972)
Corrugated facies; plains facies; mare			Domical facies	Textured ejecta	Head (1974)
Floor material			Knobby material	Hevelius formation	Howard et al. (1974)
Central basin material			Knobby basin material	Concentric facies; radial facies; smooth plains; grooved facies	Moore et al. (1974)

between these rings is made up of highly fractured, closely
spaced, smooth hills. Inferences on both the depth of excavation and the volume of the basin ejecta are necessarily dependent on the interpretation of this region.

There are two dominant theories for the formation of the
knobby or domical material (Table I). The origin has been
ascribed to initial deposition of thick ejecta and later modification by seismic activity accompanying the formation of Montes Cordillera (Head, 1974a and 1976a), or the fallback of
ejecta during the terminal stages of the cratering event (Howard et al., 1974; Moore et al., 1974). McCauley (1976) suggested that the material may represent coherent blocks excavated from beneath the 60 km seismic discontinuity at the base
of the lunar crust. There is photogeologic support for both
interpretations, and inferences depend on the mode of origin
of the outer basin rings and the size of the transient cavity.

Calculation of the volume of material ejected from Orientale based on gravity data indicates a minimum of 5.3 x 10^6 km^3
of ejecta, based on the Cordillera ring as the original crater
rim (Scott, 1974). As shown by Moore et al. (1974), this
value agrees well with the amount of radial ejecta (4.5 x 10^6
km^3) estimated from the region outside the Cordillera ring;
the transient cavity was assumed to be smaller than this ring
and about the size of the outer Rook ring (Moore et al., 1974,
p. 84).

Hodges and Wilhelms (1976) advocate an initial crater rim
at the Cordillera ring by analogy with terrestrial impact
craters. They believe that the inner rings may represent
layering in the crust, structurally deformed by rebound of the
crater floor. (This interpretation is supported in part by the
mode of formation of central peaks in craters smaller than 100
km, and peaks and rings in large craters between 100 and 300 km
in diameter, (Fig. 6). In contrast to this interpretation,
Head et al. (1975) estimate an ejecta volume of only 1 - 3 x
10^6 km^3 using the outer Rook ring as the initial crater of
excavation. Consequently, the Orientale event would not have
ejected material from deeper than 20 km.

Recent mapping of the Orientale basin indicates that the
Cordillera scarp is not as well developed on the western side
of the basin, and that the region between the Rook and Cordillera mountain chains does have radial lineations suggestive
of ejecta (McCauley, 1976). In addition, ratios of basin rings
to central peak rings also imply that the outer Rook Mountains
are the site of the original crater rim (Head, 1976a). Consequently, although Orientale is well preserved, details of its
formation are not yet entirely understood.

Correct interpretation of the structural history of the
Orientale basin is of great consequence both to highland stratigraphy of the western limb region, and to the extrapolation

of data to older, less well-preserved lunar basins. Using an
empirical relationship between ejecta thickness and crater
radius from McGetchin et al. (1973), the thickness of Orientale
ejecta at the western edge of Oceanus Procellarum would be
about 90 m if the outer Rook ring represents the original crater, and 270 m if the Cordillera ring is the original rim. The
difference becomes more pronounced closer to the basin, and at
the Cordillera ring, thickness estimates range from 0.5 to
2.1 km, depending on which basin model is used. While the
exact thickness and extent of the Orientale ejecta is a matter
of controversy, the observation that ejecta of this and other
lunar basins and craters overlie and subdue preexisting craters
is undisputed.

Light Plains and Terra Domes

     In addition to deposits of basins and craters of varying
sizes, the lunar highlands display numerous other features.
Two types of highland landforms have puzzled lunar photogeolo-
gists for a long time: (1) the light-colored, heavily cratered,
but otherwise featureless, smooth plains that cover about 4% of
the lunar surface (Howard et al., 1974); and (2) rugged, highly
textured domical structures with distinct multispectral sig-
natures.
     Prior to sampling the Cayley-type lunar light plains on
Apollo mission 16 north of Descartes crater, they were considered as most likely volcanic. Although the plains were initially interpreted by Eggleton and Marshall (1962) as smooth
facies of Imbrium basin ejecta, their distribution over the
Moon and at varying topographic levels favored the volcanic
theory. They were interpreted as possibly old marelike basalt
or more silicic lavas and/or pyroclastic materials (e.g., Milton, 1968; Wilhelms and McCauley, 1971). Free-fall or ash-flow
tuff was suggested by the similarity beyween plains and adjacent
mantled terra units (Howard and Masursky, 1968).
     As explained by Hodges et al. (1973) the volcanic interpre-
tation appeared more plausible for the following reasons:

     1. The Cayley plains are similar to mare plains in most
aspects except for a higher albedo (different composition) and
a higher crater density (longer exposure age).
     2. The plains occupied unconnected topographic lows and
therefore appear to be locally derived.
     3. Within a few plains units (e.g., the floor of the
crater Alphonsus), volcanic vents occur along linear depressions.
     4. The plains fill in and truncate_sculpture of the Im-
vrium basin and thus are assumed to be younger.

The Apollo 16 samples indicated that the light plains are predominantly impact breccias (LSPET, 1973). Because of this, numerous theories have been advanced to explain the light plains as relatively fine-grained ejecta from multiringed basins and craters:

1. *Multiple Basin Ejecta*. This theory was advanced by Hodges *et al.* (1973, p. 1), who recognized that the Moon-wide occurrence of Cayley-like plains and the apparent impact origin of the returned samples suggest a possible relation of such plains deposits to multi-ringed impact basins. The apparent contemporaneity of all the Imbrium light plains units, including those around and genetically related to the Orientale basin, suggests further that at least the top layer of these deposits may be a product of the Orientale impact. It seems probable that the total thickness of plains material at Apollo 16 comprises a sequence of deposits from multi-ringed basins, including Nectaris and Imbrium as well as Orientale.

Hodges *et al.* (1973) carried this idea further to conclude that since the seemingly best example of viscous highland volcanism (the Descartes highlands) was discredited, it was unlikely that such a process did operate on the Moon. They believed that other areas previously mapped as possible products of highland volcanism may be interpreted as products of multiring basin impacts. However, this should not include terra domes that cannot be explained in that manner. This author, among others, believes that there are several structures that are probably the product of viscous terra volcanism, such as the Mons Hansteen and the Mons Gruithuisen domes (Fig. 7).

2. *Orientale Basin Ejecta*. Chao *et al.* (1975) further refined the multiringed basin ejecta theory and concluded that most Cayley-type, Imbrian-age plains may be composed mainly of Orientale basin ejecta. They based this on the fact that stratigraphic relations and crater size-frequency distributions, and dating by erosional morphology of superposed craters, they have established that the plains are younger than the Imbrium basin and older than the maria. Therefore, this hypothesis implies that the Orientale impact struck a highland area underlain by feldspathic material and spread it over much of the Moon.

3. *Secondary Impact Ejecta*. Oberbeck *et al.* (1975) disagreed with the two hypotheses concluding that light-plains cannot be solely basin ejecta. They based their new idea on the premise that ejecta thrown beyond the continuous deposits of basins and large craters produces secondary impact craters that excavate and deposit masses of local material equal to multiples of that of the primary crater ejecta deposited at the same place. They support this interpretation by the fact that

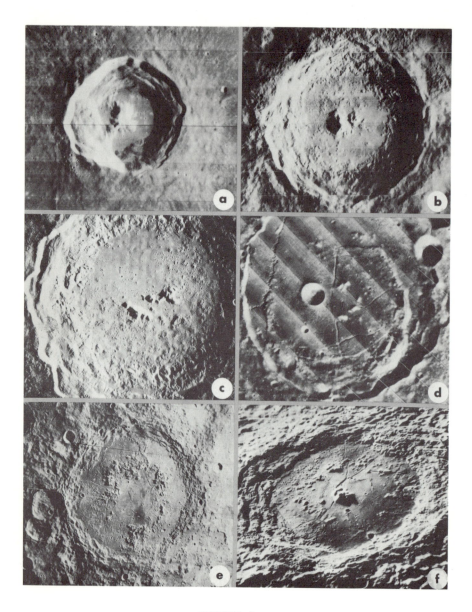

FIGURE 6.

FIGURE 6. This sequence of craters illustrates stages of central peak and inner ring development. Peak and ring morphology is in part size-dependent, and is probably affected by subsurface characteristics and impact velocity as well. (a) Crater Burg, 40 km in diameter, located in Lacus Mortis, north of Mare Serenitatis. It exhibits a simple, single peak representative of many craters of its size range (LO IV H-91); (b) Crater Tycho, 70 km in diameter, displaying a split or double peak (LO IV H-119). (c) Crater Copernicus, 90 km in diameter where the central peak is spread into large detached segments (LO IV M-151). (d) Isolated peaks within the crater Posidonius, 100 km in diameter, arranged in a small incipient ring (LO IV H-79 and H-86). (e) The basin Schrödinger, 320 km in diameter, displays a well-defined but discontinuous inner ring (LO M-8). (f) The crater Compton, 170 km in diameter, showing a central peak in addition to an inner ring (LO V M-181). This sequence may ultimately lead to multiringed basins like Orientale (see Fig. 3).

FIGURE 7. Two lunar orbiter images illustrating rugged, highly textured domes of probable (silicic) volcanic origin: left, the Mons Hansteen structure, 20 km in diameter, on the southern edge of Oceanus Procellarum (LO IV H-149); and right, Mons Gruithuisen Gamma (center, about 20 km across) and Mons Gruithuisen Delto- (right center) in northeastern Oceanus Procellarum (LO IV H-145).

several plains units have compositions similar to adjacent
highlands, but different from other plains regions.

   4. *Local Impact Ejecta*. Head (1974b) interpreted the
stratigraphy of the Cayley-like plains in the Apollo 16 site
region as a result of complex interaction of deposits of local
and regional impact events. He proposed that large 60 - 150 km
diameter craters have had a dramatic effect on the history and
petrogenesis of that region. He concluded therefore that contributions from Imbrium are minor and those from Orientale are
negligible.

   Detailed geologic mapping of the Moon does not support the
interpretation of lunar light-colored plains as the product of
one or a few impact events. In the course of mapping the east
side of the Moon (Wilhelms and El-Baz, 1977) different light-
plains units were encountered. Some of these are most likely
the products of local and regional impacts, others may be volcanic in origin. Figure 8 illustrates the geology of this
region. As explained by El-Baz and Wilhelms (1975), the
mapped geologic provinces (where a geologic unit or groups of
units related in age and origin are concentrated) include the
following (Fig. 8c):

   1. *Cratered Terra*. This province is mostly on the far
side and consists of densely packed craters. It contains
little basin materials except for the ancient, subdued rings
of the Al-Khwarizmi/King (El-Baz, 1973), Tsiolkovskij/Stark,
and Lomonosov/Fleming basins (Fig. 8b), and a few additional
short arcs of rings. The province owes its preservation to a
lack of significant modification by relatively young basins,
and its materials are the most primitive on the east side of
the Moon.

   2. *Basin Rims*. Topographically the most rugged, this province includes the multiple rings and some peripheral terrain
of several basins. It is most extensive around the Crisium,
Marginis, and Smythii basins.

   3. *Mantled Terra*. This province includes extensive tracts
of terra that appear mantled near young and old basins. Although the distinctive lineated textures usually associated with
basin ejecta have not been observed in this province, its proximity to basins suggests that it is composed mostly of degraded
basin ejecta.

   4. *Lineated Ejecta*. The five Nectarian basins in the area
are surrounded by radially lineated ejecta and clusters of
secondary impact craters. These features are diagnostic of
the impact origin of the source basins, and are most extensive
around the Humboldtianum and Nectaris basins.

   5. *Old Plains*. This province includes densely cratered,
light-colored plains in topographically low areas. Although
most plains may be of impact origin, some may be volcanic.

FIGURE 8a. Photograph of the east side of the Moon showing major surface features. Note the large area occupied by light-colored plains between the craters Lomonosov and Fleming. (Apollo 16 metric photograph 3023.)

6. *Young Plains.* The province includes light-colored plains that are less densely cratered than old plains, and fills low areas near basins and within large craters. Most occurrences are probably derived from impact melts; however, a volcanic origin cannot be excluded, particularly for the plains within the Lomonosov/Fleming basin (Fig. 8a,b).

7. *Old Mare.* The largest expanse of a relatively old mare province is in the Australe basin. In this locality the mare is overlapped by materials of the Imbrian-age craters Humboldt (27°S, 81°E) and Jenner (42°S, 96°E). This relation indicates that this province includes units that are older than most or all nearside mare materials.

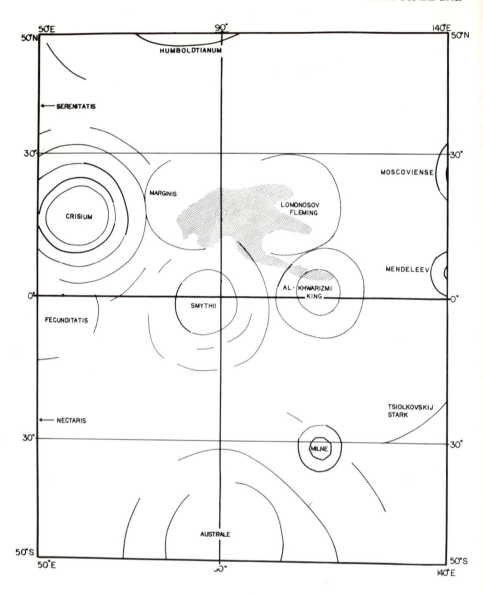

FIGURE 8b. Basin rings on the east side region of the
Moon. Solid lines indicate conspicuous arcs and dotted lines
indicate subdued rings. The stippled area centered at about
15°N, 19°E shows light-colored swirls in and northeast of
Mare Marginis (after El-Baz and Wilhelms, 1975).

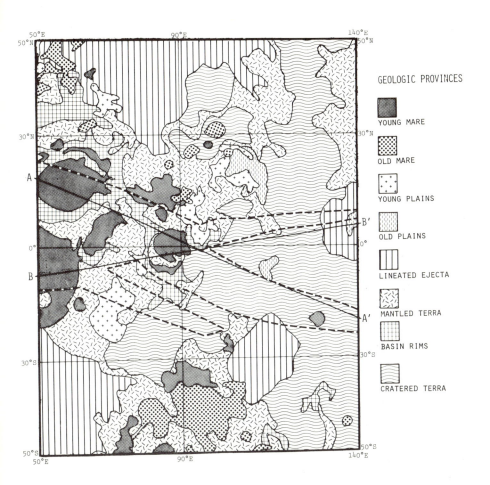

FIGURE 8c. Geologic provinces on the east side region of the Moon. Lines AA' and BB' represent tracks of laser altimeter measurements and dashed envelope represents the extent of geochemical remote-sensing on Apollo missions 15 and 16 (after El-Baz and Wilhelms, 1975).

8. *Young Mare*. The young mare units are concentrated mainly in the Crisium, Fecunditatis, and Smythii basins and the troughs between them. The crater Tsiolkovskij in the south-eastern part of the region (20°S, 129°E) and a few other craters and depressions on the far side within larger basins contain maria. A conspicuous farside patch of young mare materials not near either a basin or a large crater is centered at 27°S, 130°E (see El-Baz, 1972, pp. 48-49).

THE MARIA

Prior to the Apollo lunar missions, most investigators interpreted the dark lunar plains as volcanic (probably basaltic) lava flows. This interpretation was confirmed by analyses of lunar samples returned by Apollo 11, 12, 15, and 17 and Luna 16 (0°42'S, 56°18'E) missions.

The maria are mostly confined within the multiringed basins and nearby troughs (Fig. 9). However, these basins are older than and not genetically related to, the basalts they contain. They only provide the depressions in which the basalts rest. (Fractures that may have been produced at the time of basin formation may have provided channel ways for the later upward movement of basaltic magmas.)

As illustrated in Fig. 9, the maria constitute about 17% of the lunar crust. Head (1976b) calculates that volumetrically, the maria compose only 1% of the crustal materials. However, mare basalts (which are compositionally simpler than highland rocks), contain much information about the thermal history of the Moon and the nature of the lunar interior (Papike *et al.*, 1976).

Because of the preponderance of flow fronts and other superposition relationships in the lunar maria (El-Baz, 1974), age relationships between mare units are relatively easy to decipher. Where superposition relationships are not distinct, three methods of dating based on crater density may be used:

(1) Assuming a known flux rate of impact, the absolute crater frequency will be proportional to the age of the surface (Showmaker and Hackman, 1962).

(2) Considering the morphologically oldest superposed crater on a mare surface, a sequence of lava flooding can be deduced (Pohn and Offield, 1970).

(3) Determining the maximum diameter of craters that have been eroded such that the interior slope is less than the Sun angle. The maximum crater diameter is then converted to an equivalent diameter (D_L) of a crater that would have been eroded

to an interior slope of 1^0 by the net accumulated flux. Therefore, values of D_L became proportional to the total number of
craters on a given surface (Soderblom and Lebofsky, 1972).

Application of these methods has resulted in working out
detailed stratigraphic sequences in Mare Imbrium (Boyce and
Dial, 1975), in Oceanus Procellarum (Boyce, 1975), and in
southeastern Mare Serenitatis (Maxwell, 1977). Data on relative ages in these studies are usually correlated with albedo
and color-difference data of Whitaker (1972). For example,
Boyce and Dial (1975) recognized the following units in Mare
Imbrium and Sinus Iridum:

1. An old, low albedo unit in the southeastern corner of
the Imbrium basin.
2. An old, intermediate red unit around the edge of the
basin.
3. A young, intermediate red unit in Sinus Iridum, which
is contemporaneous with a blue unit.
4. A young, blue unit in the southwest and central parts
of the basin.

The relative age scheme of mare units can not be calibrated
by the results of age dating of returned mare samples. These
samples vary in age between 3.15 and 3.85 billion years (Tera
et al., 1974). However, the aforementioned relative age
dating techniques indicate that basalts as young as 2.5 billion
years exist in Oceanus Procellarum, within an area of the Moon
that has not yet been sampled.

Chemically, the sampled lunar mare basalts can be divided
into two groups: old (3.55 - 3.85 billion years), high-titanium
basalts of Apollo 11 and 17; and relatively young (3.15 - 3.45
billion years), low-titanium basalts of Apollo 12 and 15 and
Luna 16. According to Papike et al. (1976), these two groups
were derived from mineralogically distinct source regions.
The low-titanium basalts could have been derived from an oli-
vine-pyroxene source rock at depths ranging from 200 to 500
km, while the high-titanium basalts could have been derived
from olivine-pyroxene-ilmenite cumulates in the outer 150 km
of the Moon.

LUNAR GEOLOGIC PROVINCES

After the completion of detailed geologic mapping of the
Moon, it is now possible to divide the lunar materials into
six distinct provinces (Howard et al., 1974). As explained by
Wilhelms (1974), to best compare these to geologic provinces
on Mars, the maria are divided into young and old. The highlands or terrae are divided into four provinces according to

even

FIGURE 9a. Distribution of lunar mare materials (black
areas) on the near side of the Moon. (Base map is an equal
area projection by the National Geographic Society.)

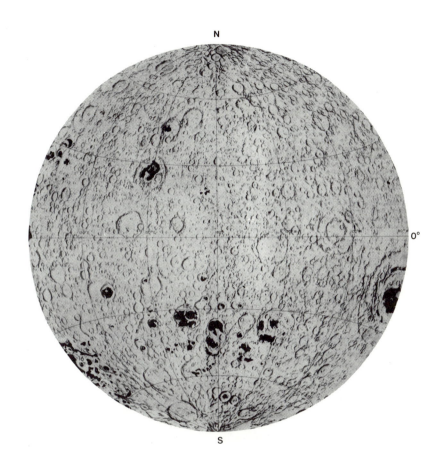

FIGURE 9b.   Distribution of lunar mare materials (black areas) on the far side of the Moon.   (Base map is an equal area projection by the National Geographic Society.)

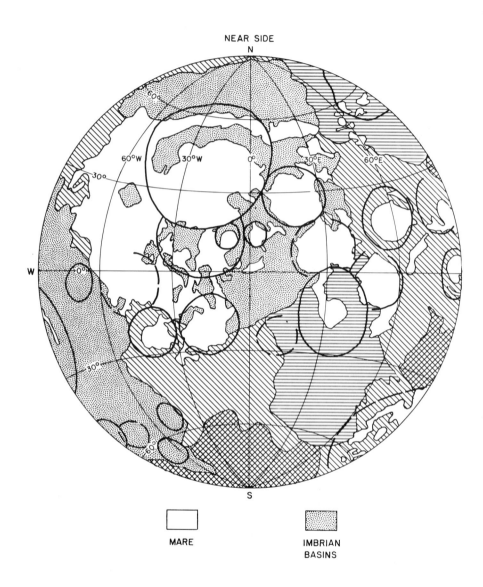

FIGURE 10a. Thematic map of major geologic provinces on
the Moon. For explanation of provinces refer to text (after
Howard et al., 1974).

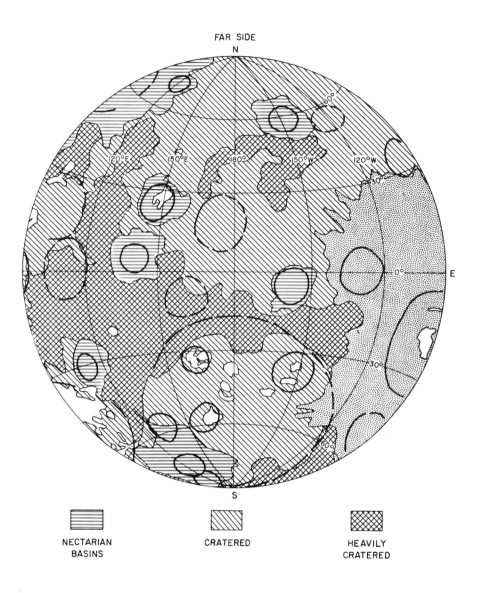

FIGURE 10b. Thematic map of major geologic provinces on
the Moon. For explanation of provinces refer to text (after
Howard et al., 1974).

the degree of development of multiringed circular basins
(Fig. 10). Wilhelms (1974) defines these provinces as follows:

1. *Heavily Cratered Provinces*. Parts of the lunar highlands show no evidence of basin ejecta blankets, although they
may contain some degraded basin rings. In this province there
is a high density of overlapping craters, probably saturated
to sizes of 200-300 km or even to the size range of basins
(Howard *et al.*, 1974).

2. *Cratered Provinces*. This province, which has fewer old
and degraded craters than the heavily cratered province,
occurs near basins and contains degraded basin rings and much
undulatory mantling material that is believed to be basin
ejecta. The basins have apparently reduced the number of
visible craters from that of the heavily cratered province.

3. *Nectarian Basin Province*. Basins of Nectarian age
(that is, late pre-Imbrian) have distinct ejecta and secondary
craters and rugged raised rings. The largest basins are Nectaris (the oldest) and Humboldtianum; the other smaller ones
are more like craters but still have the multiple rings
(usually two) that distinguish basins from craters (Hartmann
and Kuiper, 1962; Stuart-Alexander and Howard, 1970; Hartmann
and Wood, 1971) The basin materials have obliterated or obscured old craters and form surfaces on which there are fewer
craters than in the first two provinces.

4. *Imbrian Basin Province*. Young lunar multiringed impact basins have obvious similarities to impact craters, including sharply lineated ejecta blankets and satellitic clusters of fresh-appearing secondary impact craters (summaries by
Stuart-Alexander and Howard, 1970; Hartmann and Wood, 1971;
Wilhelms and McCauley, 1971; Wilhelms, 1973; Howard *et al.*,
1974). Materials of the two basins of Imbrian age, Imbrium and
Orientale, constitute the province that is most obviously of
basin origin. The basin ejecta blankets have covered pre-Imbrian and early Imbrian craters and have thereby given rise to
the lowest crater densities on the lunar terrae.

5. *Old Mare Province and Light Plains Material*. As stated
above, in the Mare Australe region of the Moon (Fig. 8c), mare
material is pitted by the secondary impacts of the craters
Humboldt and Jenner, which themselves contain mare material
(Stuart-Alexander and Howard, 1970, p. 451; see also El-Baz and
Wilhelms, 1975; Wilhelms and El-Baz, 1977). These craters are
of Imbrian age. Therefore the Australe mare material is older
than the near side maria. A few other patches of the old
mare material are also present elsewhere on the far side.

A similar but lighter-colored lunar plains unit was formed
at about the same time as the old mare material. The two units
have a similar crater density, but the light plains appear in
places to be overlain by the dark. The light plains are no-

where sufficiently concentrated to be considered a province at
this scale. The greatest concentrations occur near the Orientale and Imbrium basins and are included in the Imbrian basin
province, in accord with the interpretation that these plains
are mostly basin ejecta (Howard *et al.*, 1974).

   6. *Young Mare Province*. All of the maria on the lunar
near side and some mare patches on the far side are assigned
to this province. The range of ages is from Imbrian to
Eratosthenian, or from 3.85 to 3.15 billion years ago as so
far sampled and dated. This period is 15% of the Moon's
history but about 50% of the part of the history that included
major surface changes, for the subsequent impacts did not
severely obscure the maria. The era represented by the maria
was itself much less active in terms of impact events than the
preceding 50%, for in the earlier epoch the terra materials
were repeatedly reworked and redistributed by a rain of impacting objects.

SUMMARY

   Utilization and testing of the concepts of stratigraphy results in a reasonable account of lunar surface history. The
evolution of the lunar crust is schematically shown in Table II,
in which correlations are made between the major geologic provinces, the relative-age scheme of lunar surface materials, and
the absolute ages as deduced mostly from returned lunar samples.
   From this and preceding discussions, the evolution of the
lunar crust can be divided into five major stages:
   1. *Formation of the Original Crust*. Little is known about
the early period of lunar crustal formation. Based on model
ages of lunar soils, the Moon is assumed to have formed 4.6
billion years ago. This age is equal to that of meteorites
and also to the one assumed for the solar system. Therefore,
it is reasonable to assume that it represents the condensation
age of the solar nebula. As the condensed material formed the
Sun and the planets, some of it gathered to form the Moon. As
the moonlet grew, it was repeatedly struck by a large amount
of infalling debris. This probably resulted in melting the
upper 100 - 300 km of the Moon by accretional energy. The
melting resulted in a Moon-wide differentiation to form the
feldspar-rich crustal rocks (anorthositic gabbros and gabbroic
anorthosites).
   2. *Heavy Cratering Episode*. As the amount of infalling
objects decreased, the differentiated crust started cooling.
When the crust solidified, it started preserving the scars of
collisions with large and small planetesimals. Many basins and
numerous craters pockmarked the surface over a long period of

time, where relatively new features obliterated or subdued older
ones. This episode probably started as early as 4.55 billion
years ago and continued to about 3.9 billion years ago. Remnants of the oldest impact scars are probably no longer visible.

TABLE II.   *Correlation of Nomenclature of the Orientale
Basin and Related Materials.*

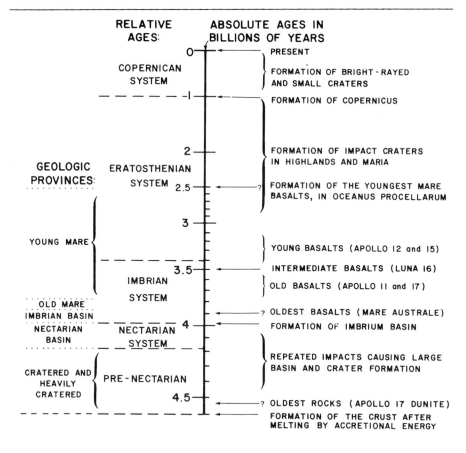

The more recent impacts with discernible remains may be classed
as early (including basins like Al-Khwarizmi/King, Australe,
and South Pole/Aitken), middle (including Smythii, Nectaris,
and Crisium), and late (including the youngest basins, Imbrium
and Orientale, which probably formed at 3.95 and about 3.9
billion years ago, respectively).

    3. *Formation of the Oldest Plains.* In many parts of the
Moon, both light-colored (highlandlike) and dark-colored (marelike) plains formed contemporaneously with and after the formation of the youngest lunar multiringed basins. These plains
probably span a period from 4.0 to 3.85 billion years ago.
Most light-colored plains are genetically related to basin-
forming impact events, although some of them may be volcanic
in origin. The old dark-colored plains are probably composed
of basalts.

    4. *Major Basaltic Flooding.* Differentiated basaltic mag-
mas were probably originated by melting due to heat generated
by radioactive decay beneath the lunar crust. These magmas
probably made their way to the surface via impact-induced
fractures in the crust. Lavas flowed on the surface and
accumulated in low areas, particularly within the multiringed
basins. From the absolute age dating of returned lunar samples,
this major episode of basaltic lava flooding may be divided
into three distinct phases: (1) old, between 3.85 and 3.5
billion years ago (Apollo 11 and 17 rocks); (2) intermediate,
between 3.5 and 3.4 billion years ago (Luna 16 rocks); and
(3) young, between 3.4 and 3.15 billion years ago (Apollo 12
and 15 rocks). A fourth phase, as young as 2.5 billion years,
is suggested by crater age dating in Oceanus Procellarum.

    5. *Stabilization of the Crust.* Near the end of the main
basaltic flooding stage, the major features of the Moon's
surface had been formed. Events of only local significance en-
sued. Among these are the impacts that created many of the
prominent lunar craters that are smaller than 100 km in diameter, e.g., Eratosthenes (about 3.2 billion years ago), Copernicus (about 1.0 billion years ago), and Tycho (70 - 95 million
years ago). Dating of returned lunar samples also provided
data on the ages of craters in the Apollo landing sites. For
example, Shorty Crater at the Apollo 17 site (20 - 30 million
years ago), and South Ray crater at the Apollo 16 site (only
2 million years ago). Smaller meteoroid impacts continue to
locally modify lunar surface stratigraphy to this day.

ACKNOWLEDGMENTS

The author is indebted to T. A. Maxwell, P. L. Strain, and
R. W. Wolfe for reviewing this paper. Thanks are due to D. J.
Hennen and J. E. Murphy for typing and proofing the manuscript,
and to A. W. Gifford for preparing the illustrations.

REFERENCES

Baldwin, R. B. (1963). *"The Measure of the Moon."* Univ. of
    Chicago Press, Chicago, Illinois.
Baldwin, R. B. (1972). The tsunami model of the origin of ring
    structures concentric with large lunar craters, *Phys. Earth
    Plan. Interiors 6,* 327-339.
Boyce, J. M. (1975). Chronology of the major flow units in the
    western nearside maria, *in Conf. Origins of Mare Basalt,*
    LSI, Houston, Texas, pp. 11-14.
Boyce, J. M., and Dial, A. L. (1975). Relative ages of flow
    units in Mare Imbrium and Sinus Iridum, *Proc. Lunar Sci.
    Conf. 6th 3,* 2585-2595.
Chao, E. C. T., Hodges, C. A., Boyce, T. M., and Soderblom,
    L. A. (1975). Origin of lunar light plains, *J. Res. U. S.
    Geol. Survey 3,* 379-392.
Dence, M. R. (1972). The nature and significance of terrestrial
    impact structures, *Internat. Geol. Cong-. Rept. Sess. 24th,*
    pp. 77-89.
Eggleton, R. E. (1964). Preliminary geology of the Riphaeus
    quandrangle of the Moon and definition of the Fra Mauro
    Formation, *in Astrogeol. Stud. Ann. Prog. Rept. (1962-63),*
    Pt. A, pp. 43-46.
Eggleton, R. E., and Marshall, C. H. (1962). Notes on the
    Apenninian series and pre-Imbrian stratigraphy in the vicin-
    ity of Mare Humorum and Mare Nubium, *in Astrogeol. Studies
    Semiann. Prog. Rept., Pt. A,* pp. 99-109.
El-Baz, F. (1972). New geological findings in Apollo 15 lunar
    orbital photography, *Proc. Lunar Sci. Conf. 3rd 1,* 39-61.
El-Baz, F. (1973). Al-Khwarizimi: A new-found basin on the
    lunar far side, *Science 180,* 1173-1176.
El-Baz, F. (1974). Surface geology of the Moon, *Ann. Rev.
    Astron. Astrophys. 12,* 135-165.
El-Baz, F. (1975). The Moon after Apollo, *Icarus 25,* 495-537.
El-Baz, F., and Wilhelms, D. E. (1975). Photogeological, geo-
    physical, and geochemical data on the east side of the moon,
    Proc. Lunar Sci. Conf. 6th 3, 2721-2738.

Fielder, G. (1965). "*Lunar Geology,*" Butterworth, London.

Gault, D. E. (1970). Saturation and equilibrium conditions for impact cratering on the lunar surface: criteria and implications, *Radio Sci. 5,* 273-291.

Gault, D. E., Quaide, W. L., and Oberbeck, V. R. (1968). Impact cratering mechanics and structures, *in* (B. M. French and N. M. Short, eds.), "*Shock Metamorphism of Natural Materials*" pp. 87-99.

Gilbert, G. K. (1893). The Moon's face, a study of the origin of its features, *Phil. Soc. Washington Bull. 12,* 241-292.

Gilbert, G. K. (1896). The origin of hypotheses, illustrated by the discussion of a topographic problem, *Science 3,* 1-12.

Green, J. (1962). The geosciences applied to lunar exploration, *in* (Z. Kopal and Z. K. Mikhailov, eds.), "*The Moon*", pp. 160-257. Academic Press, London.

Hartmann, W. K., and Kuiper, G. P. (1962). Concentric structures surrounding lunar basins, *Comm. Lunar Planet. Lab. 1,* 51-66.

Hartmann, W. K., and Wood, C. A. (1971). Moon: Origin and evolution of multi-ring basins, *The Moon 3,* 3-78.

Head, J. W. (1974a). Orientale multi-ringed basin interior and implications for the petrogenesis of lunar highland samples, *The Moon 11,* 327-356.

Head, J. W. (1976a). Origin of rings in lunar multi-ringed basins. Abstracts from *Symp. Plan. Cratering Mechanics,* LSI, Houston, Texas, pp. 47-49.

Head, J. W. (1976b). Lunar volcanism in space and time, *Rev. Geophys. Space Phys. 14,* 265-300.

Head, J. W., Settle, M., and Stein, R. S. (1975). Volume of materials ejected from major lunar basins and implications for the depth of excavation of lunar samples, *Proc. Lunar Sci. Conf. 6th 3,* 2805-2829.

Hodges, C. A., Muehlberger, W. R., and Ulrich, G. E. (1973). Geologic setting of Apollo 16, *Proc. Lunar Sci. Conf. 4th 1,* 1-25.

Hodges, C. A., and Wilhelms, D. E. (1976a). Formation of concentric basin rings, Abstracts from *Symp. Plan. Cratering Mechanics,* LSI, Houston, Texas, pp. 53-55.

Hodges, C. A., and Wilhelms, D. E. (1976b). Formation of lunar basin rings, *25th Internat. Geol. Cong.,* Sydney, Australia.

Howard, K. A., and Masursky, H. (1968). Geologic map of the Ptolemaeus quadrangle of the Moon, *USGS Misc. Geol. Inv. Map I-566.*

Howard, K. A., Wilhelms, D. E., and Scott, D. H. (1974). Lunar basin formation and highland stratigraphy, *Rev. Geophys. Space Phys. 12,* 309-327.

King, E. A. (1976). "*Space geology, an introduction.*" Wiley, New York.

Lindsay, J. F. (1976). Lunar stratigraphy and sedimentology, in (Z. Kopal and A. G. W. Cameron, eds.), "Developments in Solar System and Space Science," Vol. 3 Elsevier, Amsterdam.

LSPET: Lunar Sample (Apollo 16) Preliminary Examination Team (1973). The Apollo 16 lunar samples: petrographic and chemical description, Science 179, 23-34.

Maxwell, T. A. (1977). Stratigraphy and tectonics of southeastern Serenitatis. PhD. Dissertation, Univ. of Utah.

McCauley, J. F. (1967). The nature of the lunar surface as determined by systematic geologic mapping, in "Mantles of the Earth and Terrestrial Planets" (S. K. Runcorn, ed.), pp. 431-460. Wiley, New York.

McCauley, J. F. (1968). Geologic results from the lunar precursor probes, AIAA J. 6, 1991-1996.

McCauley, J. F. (1976). Orientale and Caloris. Abstracts from the Conf. Comparisons of Mercury and the Moon, LSI, Houston, Texas, p. 24.

McGetchin, T. R., Settle, M., and Head, J. W. (1973). Radial thickness variation in impact crater ejecta: Implications from lunar basin deposits, Earth Plan. Sci. Lett. 20, 226. 236.

Milton, D. J. (1968). Geologic map of the Theophilus quadrangle of the Moon, USGS Map I-546.

Moore, H. J., Hodges, C. A., and Scott, D. H. (1974). Multiringed basins--illustrated by Orientale and associated features, Proc. Lunar Sci. Conf. 5th 1, 71-100.

Mutch, T. A. (1972). "Geology of the Moon: a Stratigraphic View," revised ed. Princeton Univ. Press, Princeton, New Jersey.

Oberbeck V. R., Horz, F., Morrison, R. H., Quaide, W. L., and Gault, D. E. (1975). On the origin of the lunar smooth plains, The Moon 12, 19-54.

Papike, J. J., Hodges, F. N., Bence, A. E., Cameron, M., and Rhodes, J. M. (1976). Mare basalts: Crystal chemistry, mineralogy, and petrology, Rev. Geophys. Space Phys. 14, 475-540.

Pohn, H. A., and Offield, T. W. (1970). Lunar crater morphology and relative age determination of lunar geologic units--Part 1, USGS Prof. pap. 700-C, 153-162.

Scott, D. H. (1974). The geologic significance of some lunar gravity anomalies, Proc. Lunar Sci. Conf. 5th 3, 3025-3036.

Shoemaker, E. M. (1962). Interpretation of lunar craters, in "Physics and Astronomy of the Moon" (Z. Kopal, Ed.), pp. 283-359.

Shoemaker, E. M., and Hackman, R. J. (1962). Stratigraphic basis for a lunar time scale, in The Moon--Symposium 14--IAU" (Z. Kopal and Z. K. Mikhailov, eds.), pp. 289-300. Academic Press, New York.

Short, N. M. (1975). *"Planetary Geology."* Prentice-Hall Inc.,
    Englewood Cliffs, New Jersey.
Soderblom, L. A., and Lebofsky, L. A. (1972). Technique for
    rapid determination of relative ages of lunar areas from
    orbital photography, *JGR 77*, 279-296.
Spurr, J. E. (1944). The Imbrian plain region of the Moon, *in*
    "Geology Applied to Selenology," Vol. 1. Science Press,
    Lancaster, Pennsylvania.
Spurr, J. E. (1945). The features of the Moon, *"Geology Applied
    to Selenology,"* Vol. II. Science Press, Lancaster, Pennsyl-
    vania.
Stuart-Alexander, D. E. (1976). Geologic map of the central
    far side of the Moon, *USGS Interagency Rept.: Astrogeology
    79.*
Stuart-Alexander, D. E., and Howard, K. A. (1970). Lunar maria
    and circular basins--a review, *Icarus 12*, 440-456.
Stuart-Alexander, D. E., and Wilhelms, D. E. (1974). Nectarian
    system, a new lunar stratigraphic name, *U.S. Geol. Surv.
    Jour. Res. 3*, 53-58.
Taylor, S. R. (1975). *"Lunar Science: A Post-Apollo View."*
    Pergamon, New York.
Tera, F., Papanastassiou, D. A., and Wasserburg, G. J. (1974).
    The lunar time scale and a summary of isotopic evidence for
    a terminal lunar cataclysm, *in "Lunar Science IV,* LSI,
    Houston, Texas, pp. 792-794.
Whitaker, E. A. (1972). Lunar color boundaries and their re-
    lationships to topographic features: A preliminary survey,
    The Moon 4, 348-355.
Wilhelms, D. E. (1965). Fra Mauro and Cayley Formations in
    Mare Vaporum and Julius Caesar quadrangles, *in Astrogeol.
    Stud. Ann. Rept., Pt. A*, pp. 13-28.
Wilhelms, D. E. (1970). Summary of lunar stratigraphy--Tele-
    scopic observations, *U. S. Geol. Surv. Prof. Pap. 599-F.*
Wilhelms, D. E. (1973). Comparison of martian and lunar multi-
    ringed circular basins, *JGR 78*, 4084-4095.
Wilhelms, D. E. (1974). Comparison of martian and lunar geo-
    logic provinces, *JGR 79*, 3934-3941.
Wilhelms, D. E., and Davis, D. E. (1971). Two former faces of
    the Moon, *Icarus 15*, 368-372.
Wilhelms, D. E., and El-Baz, (1977). Geologic map of the east
    side of the Moon, *U. S. Geol. Surv. Map I-948.* In press.
Wilhelms, D. E., and McCauley, J. F. (1971). Geologic map of
    the near side of the Moon, *U. S. Geol. Surv. Map I-703.*

7

ATMOSPHERIC EVOLUTION
ON THE INNER PLANETS

James C. G. Walker

National Astronomy and Ionosphere Center
Arecibo Observatory, Arecibo, Puerto Rico

The very different masses of the atmospheres of Venus, Earth, and Mars can be attributed to different rates of evolution of internal temperatures and patterns of tectonic activity on the different planets. According to this hypothesis, the inner planets all formed with hot interiors and most of their volatiles at the surface. The volatiles have gradually returned to the solid phase as the interiors of the planets have cooled. The rate of cooling has been greater on Mars and slower on Venus than on Earth. Internal temperatures on Mars are now so low that most of the water and carbon dioxide originally released to the atmosphere has returned to the solid phase. Most of the carbon dioxide on Earth has also returned to the solid phase, but internal temperatures are too high to have yet permitted the absorption of much water. Temperatures within Venus are still higher, preventing the incorporation of even carbon dioxide into the solid phase.

Originally, high surface temperatures on Venus were the result of the runaway greenhouse effect caused by the relative proximity of the planet to the sun. The original atmosphere contained abundant water vapor. Because an atmosphere with abundant water lacks a cold trap to concentrate the water at low altitudes, the water on Venus was soon destroyed by photolysis, followed by the escape of hydrogen to space and the reaction of oxygen with the surface.

The interior of Mars was hotter in the past and surface volatiles were much more abundant than they are now. While water and carbon dioxide have been removed from the atmosphere by weathering, nitrogen has been lost to space as a result of the escape of hot atoms produced by exothermic chemical reactions occurring in the upper atmosphere.

141

INTRODUCTION

Earth has an atmosphere of nitrogen and oxygen with a surface pressure of 1 bar. Venus has an atmosphere about 100
times as massive composed almost entirely of carbon dioxide.
Mars also has a carbon dioxide atmosphere, but one that is very
tenuous; the surface pressure is only 6 mbar. Do these differences reflect markedly different initial volatile complements
on the different planets or are they simply the result of
evolutionary processes acting at different rates to change volatile complements that were initially similar? My belief is
that evolution rather than initial conditions underlies the
differences between the modern atmospheres of the inner planets.
This chapter outlines, in qualitative and speculative terms,
the evolutionary processes that I have in mind.

It is no secret, of course, that Earth's atmosphere is in
no way representative of Earth's volatile content. If the crust
of the Earth were to release to the atmosphere, as carbon dioxide, all of the carbon it contains, we would have an atmosphere
very similar in composition to the atmosphere of Venus (Sagan,
1962; Ronov and Yaroshevskiy, 1967, 1969). If the terrestrial
oceans were to evaporate, the atmosphere would contain, in
addition, about 260 bar of water vapor. There is, moreover,
evidence for water and carbon dioxide in the upper mantle of
the earth (Roedder, 1965; Kennedy and Nordlie, 1968; Lambert and
Wyllie, 1970; Mitchell and Crocket, 1971; Dawson, 1971; Green,
1972; Perry and Tan, 1972; MacGregor and Basu, 1974), but the
amounts are not known. The differences between Venus and
Earth are not so great after all. Venus has a high surface temperature, about $750^{\circ}K$ compared with Earth's $290^{\circ}K$. Most of the
carbon dioxide on Venus is in the atmosphere rather than in the
solid phase. The atmosphere of Venus presently contains less
than 5×10^{-4} times as much water as the terrestrial ocean
(Pollack and Morrison, 1970; Janssen et al., 1973). These are
the differences between the atmospheres of Earth and Venus
that any theory of atmospheric evolution must try to explain.

The average surface temperature of Mars is $210^{\circ}K$, sufficiently low to permit substantial amounts of water and carbon
dioxide to freeze or be adsorbed into the regolith and polar
cap (Leighton and Murray, 1966; Pollack et al., 1970a,b; Fanale
and Cannon, 1971, 1974; Murray and Malin, 1973; Houck et al.,
1973; Malin, 1974; Huguenin, 1974, 1976a,b; Ingersoll, 1974;
Dobrovolskis and Ingersoll, 1975; Dzurisin and Ingersoll, 1975;
Flasar and Goody, 1976). It seems unlikely, however, that an
Earth-like complement of carbon dioxide and water can be contained in these reservoirs. Relative to Earth and Venus, therefore, the surface layers of Mars are deficient in carbon diox-

ide and water. There is also a marked deficiency in nitrogen (Dalgarno and McElroy, 1970), which is not subject to incorporation into the solid phase, even at Martian temperatures.

Now I shall suggest how these differences between the atmospheres of Earth, Venus, and Mars may have arisen as a result of evolutionary processes operating over the 4.6 billion years that have elapsed since the planets were formed.

PLANETARY ORIGINS

Many lines of evidence indicate that the inner planets accreted rapidly, achieving internal temperatures close to melting temperatures by the time they were fully formed (Hanks and Anderson, 1969; Turekian and Clark, 1969, 1975; Fanale, 1971a,b; Clark et al., 1972; Cameron, 1973; Ozima, 1973; Hills, 1973; Arrhenius et al., 1974; Siever, 1974; Tozer, 1974; Johnston et al., 1974; Weidenschilling, 1974, 1976; Harris and Kaula, 1975; Kaula, 1975; Kaula and Bigeleisen, 1975; Hartmann, 1976; Smith, 1976; Rama Murthy, 1976; Walker, 1976a). I shall therefore assume that temperatures and temperature gradients within the inner planets have been decreasing since the beginning of geological history as a result of the dissipation of accretional and gravitational heat and the decay of radioactive heat sources.

Associated with the hot origin of the planets was the segregation, during the course of planetary growth, of metallic iron into central cores overlain by silicate mantles essentially free of metallic iron. There has been little change in the oxidation state of the mantles since this differentiation occurred. As a result, the oxidation state of the gases released to the atmospheres of the planets has changed little since the planets were formed. Throughout nearly all of geological history the gases released from the solid phases have been similar, in oxidation state, to the gases released from modern terrestrial volcanoes. Water has been much more abundant than hydrogen, carbon dioxide has been much more abundant than carbon monoxide, and nitrogen has been much more abundant than ammonia (Heald et al., 1963; White and Waring, 1963; Holland, 1962, 1964; Nordlie, 1968, 1972; Fanale, 1971b; Cruikshank et al., 1973). The inner planets, therefore, have not had highly reducing atmospheres since the time of their formation (Walker, 1976b).

TECTONIC EVOLUTION

    As their interiors have cooled, the planets have all pro-
gressed through similar states (see Fig. 1) in the growth and
development of crust and lithosphere and in the evolution of
tectonic patterns (Kaula, 1975). Compared to Earth, however,

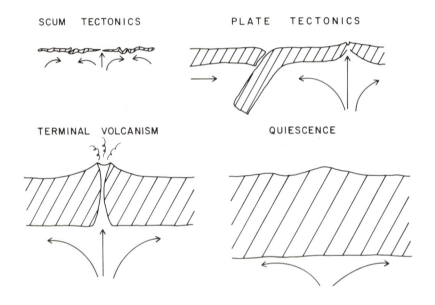

FIGURE 1. *Successive stages in the evolution of the lithosphere (cross-hatched) as temperatures and temperature gradients within a planet decrease. The arrows represent convective
motions in the underlying mantle.*

the rate of evolution has been greater for small planets like
Mars, because they cool quicker, and it has been slower for
Venus, where the atmosphere has maintained an unusually high
surface temperature.
    Initially the layer of cold, strong rock at the surface of
a planet was too thin to allow either large-scale plate motion
or substantial variation in surface elevation. During this
time the Earth may have been nearly covered by ocean overlying
a fairly uniform, thin layer of sialic crust (Hargraves, 1976).
Any tendency of sialic material to accumulate into layers of
continental thickness was counteracted by melting and redistribution of the sialic roots of the incipient continents. The
thin lithospheric shell was repeatedly disrupted by convection

of the underlying mantle. Slabs and folds of crust may have
emerged from the sea to be eroded. Crustal material, including
the products of erosion, was constantly being engulfed by the
mantle and reincorporated into it (Sutton, 1976). This has been
called the era of scum tectonics (Lambert, 1976).

As the planets cooled, the lithospheres became thicker and
stronger (Engel et al., 1974; Sutton and Watson, 1974; Shaw,
1976; Veizer, 1976). Scum tectonics was gradually replaced by
plate tectonics. On Earth, low-density sial accumulated to
form continents, while oceanic crust developed between the continents. The lithosphere was now too thick to be engulfed
piecemeal by the mantle, but it was strong enough to undergo
subduction, and so the cycling of material between the surface
and the interior continued (Armstrong, 1968; Armstrong and
Hein, 1973; Meadows, 1973; Walker, 1977a).

Plate tectonics dies away as a planet continues to cool
and its lithosphere continues to become thicker and stronger.
In the era of terminal volcanism (Kaula, 1975) there is no plate
motion but still enough internal heat to permit magmatism to
break through the lithosphere from time to time, producing
volcanic piles. There is little opportunity for crustal material to be returned to the mantle, and so the interior of
the planet may become depleted in volatiles.

The final era is one of quiescence, in which even the release of material from the interior of the planet is prevented
by the thickness and strength of the lithosphere.

It is likely that all Earth-like planets pass through these
various stages of evolution, but as already noted, they do so at
different rates. The Moon and possibly Mercury are already
quiescent (Kumar, 1976). Mars may be in the state of terminal
volcanism. Earth is undergoing plate tectonics and Venus may
have progressed even less far than Earth.

PARTITIONING OF VOLATILES BETWEEN THE ATMOSPHERE AND THE SOLID
PHASE

The evolution of internal temperature and tectonic activity
on a planet is important for the evolution of the atmosphere
because it affects the capacity of the solid phase of the planet to incorporate volatiles (meaning carbon dioxide and water).
Volatiles are removed from the atmosphere and incorporated into
the solid phase by weathering reactions, of which the following
is a schematic example:

$$CaAl_2Si_2O_8 + CO_2 + 2H_2O \rightarrow CaCO_3 + Al_2Si_2O_5(OH)_4 \qquad (1)$$

anorthite   calcite   kaolinite

This is not an irreversible process, however. When the products
of weathering are raised to sufficiently high temperatures by
burial, subduction, or tectonic activity, the volatiles are
released and can return to the atmosphere via volcanoes and
hot springs or simply by upward migration. Examples of reactions that release volatiles from the products of weathering
are

$$Al_2Si_2O_5(OH)_4 \rightarrow Al_2O_3 + 2SiO_2 + 2H_2O \tag{2}$$
Kaolinite   corundum   quartz

$$CaCO_3 + SiO_2 \rightarrow CaSiO_3 + CO_2 \tag{3}$$
calcite   quartz   wollastonite

Since high temperatures drive volatiles out of the solid
phase of a planet, the partitioning of volatiles between the
atmosphere and the solid phase depends on temperatures and
temperature gradients within the planet. The effect is illustrated schematically in Fig. 2. Early in Earth's history, at
time t_1, the temperature gradient within the planet was large,
and carbonates and hydrated minerals could survive only close
to the surface. At a later time t_2, when the temperature gradient had fallen substantially, volatile-containing minerals
were stable to greater depths (Holland, 1976), and the volatile
content of the solid phase had increased at the expense of
the atmosphere and ocean.

According to this argument, Earth's water and carbon dioxide
are gradually disappearing into the interior as the planet cools
down. Because of the different stabilities of carbonate and
hydrated minerals, the process has gone much further for carbon
dioxide, which is already largely in the solid phase, than for
water, which is still largely in the ocean (Turekian and Clark,
1975). It seems clear that the carbon dioxide content of the
atmosphere and ocean must have been larger in the past and that
it will fall in the future as the Earth continues to cool. It
also seems possible that the solid earth will begin to extract
significant quantities of water from the ocean at some time in
the future provided hydrated minerals continue to form at the
surface.

Application of this concept to Venus, Earth, and Mars is
illustrated schematically in Fig. 3. The surface temperature
of Venus is too high to permit either carbon dioxide or water
vapor to react with rocks at the prevailing pressures (Orville, 1974). Therefore, essentially all of the volatiles on
Venus are already in the atmosphere and none are being released from the solid phase of the planet at the present time
(Walker *et al.*, 1970). On Mars, the low surface temperature
and presumed low temperature gradient provides the solid

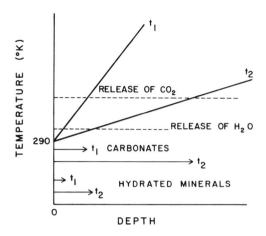

FIGURE 2. Schematic representation of the capacity of the solid Earth to contain water and carbon dioxide. The temperature gradient is smaller at time t_2 than at t_1, permitting carbonate and hydrated minerals to survive to greater depths.

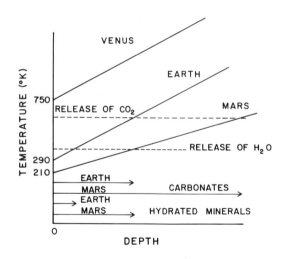

FIGURE 3. Schematic representation of the capacities for water and carbon dioxide of the solid phases of Venus, Earth, and Mars. Internal temperatures on Venus are too high to permit the retention of volatiles within the planet. Lower internal temperatures permit carbonate and hydrated minerals to survive to greater depths on Mars than on Earth.

phase with a much greater capacity for volatiles than on Earth.
Mars may provide an indication of what the Earth will be like
if the day ever arrives when most of the water follows the
carbon dioxide back into the solid phase of the planet.

This concept should not be taken to imply that the atmospheric partial pressure of a given volatile is determined by a
strict equilibrium between atmosphere and solid phase, dependent
only on the temperature profile within the solid phase. Kinetic factors involving the rates of weathering and of volcanic
release of gas are important. If atmospheric composition on
a tectonically active planet is to remain stable for geologically long periods of time, there must be a balance between the
rate at which volatiles are released from the solid phase and
the rate at which they are removed from the atmosphere by
weathering. On Earth, for example, the carbon dioxide partial
pressure is determined by the requirement that the rate at
which carbon dioxide is removed from the atmosphere by weathering of silicate minerals to form carbonate minerals be equal to
the rate at which carbon dioxide is added to the atmosphere by
volcanoes (Broecker, 1971; Walker, 1977b). The weathering rate
depends on the carbon dioxide partial pressure while the rate
of volcanic emission does not. As the capacity of the solid
phase for volatiles increases, the rate of volcanic emission
decreases, permitting a decrease in the carbon dioxide partial
pressure necessary to maintain an equal rate of weathering.
Fig. 4 illustrates the cycle schematically.

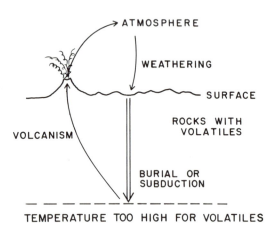

FIGURE 4.   The cycle of volcanism, weathering, and burial
that controls the volatile abundance at the surface of a planet.
Equilibrium exists when the rates of volcanism and weathering
are equal.   As the interior of the planet cools the volcanic
rate decreases; the atmospheric partial pressure must decrease
also to reduce the weathering rate.

Kinetic factors may explain the fact that the carbon dioxide pressure on Mars is greater than that on Earth in spite of the presumed greater capacity of the solid phase of Mars to contain carbon dioxide. Weathering rates may be low on Mars because of the low surface temperature and the absence of efficient mechanisms of erosion.

In summary, it does not appear to be necessary to invoke differences in initial volatile contents or different degassing histories in order to account for the different masses of the atmospheres of Venus, Earth, and Mars. The different atmospheric masses may well be direct consequences of different internal temperature profiles.

So much for general considerations. It remains to consider the special circumstances responsible for the absence of water on Venus and nitrogen on Mars as well as the high surface temperature of Venus. The special circumstance responsible for the abundance of oxygen in the terrestrial atmosphere is too well known to require discussion here.

VENUS

It is not likely that water has always been absent on Venus. The argument in favor of its abundance is due to Turekian (Turekian and Clark, 1975) and has been presented in Walker et al. (1970) and Walker (1975). It is briefly summarized here.

Theoretical studies of the primitive solar nebula indicate that carbon condensed in the form of hydrocarbons (Lewis, 1972; Grossman, 1972; Grossman and Larimer, 1974). Studies of meteorites reinforce this conclusion. Elemental carbon and carbonate minerals are rare compared with hydrocarbons (cf. Hayes, 1967). Evidently the carbon that is now present as carbon dioxide in the atmosphere of Venus (and Earth and Mars) was incorporated into the planet as hydrocarbons. These hydrocarbons were oxidized by reactions in the crust and upper mantle of the type

$$CH_2 + 3Fe_3O_4 \rightarrow CO_2 + H_2O + 9FeO \tag{4}$$

producing water as well as carbon dioxide. Corresponding to the present carbon dioxide content of the atmosphere is therefore a lower limit of $2.2 \cdot x\ 10^{23}$ gm of water, enough to yield a partial pressure at the surface of 41 bar. The Earth has six times as much water in its ocean, presumably contributed by water of hydration, but water of hydration may or may not have been incorporated into Venus (Lewis, 1972, 1974).

We conclude that Venus at one time had abundant water, which
it has lost. The special circumstance responsible for the loss
is called the runaway greenhouse effect (Ingersoll, 1969).

The release of water vapor and carbon dioxide from the interior of a planet leads to an increase in surface temperature
as a result of the greenhouse effect (cf. Goody and Walker,
1972). Approximate theoretical temperatures are shown in Fig.
5 as functions of water vapor partial pressure for Venus, Earth,
and Mars (Rasool and DeBergh, 1970). Temperatures on Mars and
Earth intersect the saturated vapor curve for water, leading to
condensation or freezing and an end to the accumulation of
water in the atmosphere. Venus, however, starts at a higher
temperature because it is closer to the sun, and its surface
temperature increases with increasing water vapor pressure, because of the greenhouse effect, more rapidly than the saturated
vapor pressure. All of the water released from the solid phase
of Venus therefore remains in the atmosphere and high surface
temperatures result (Sagan, 1960; Gold, 1964).

The essential element in the history of the Venus atmosphere
has been described by Ingersoll (1969). In an atmosphere with
abundant water vapor (a mixing ratio exceeding about 0.1),
temperature and water vapor mixing ratio decrease very slowly
with height. The infared opacity of water vapor ensures that
the vertical transport of heat through the lower reaches of
such an atmosphere is accomplished by convection. Latent heat
released by water vapor that condenses in upward moving air
counteracts the temperature decrease caused by expansion of the
air. In the nearly isothermal atmosphere that results, the
decline in saturated vapor pressure with altitude is slow,
causing little decrease with altitude in the water vapor mixing ratio.

As shown in Fig. 6, the situation is quite different in an
atmosphere, like that of the Earth, where water vapor is a
minor constituent. Even at the levels where condensation first
begins in such an atmosphere, there is not enough water vapor
(not enough latent heat) to offset the decline in temperature
caused by the expansion of upward-moving air. As temperature
falls, the capacity of the air for water vapor falls too, and
the water vapor mixing ratio decreases rapidly with height once
condensation has begun. Water vapor is therefore trapped at
low altitudes in the terrestrial atmosphere, but there would
have been no such cold trap in the primitive wet atmosphere of
Venus.

Hunten (1973) has shown that the rate at which water is lost
from a planet as a result of photolysis of water vapor followed
by escape of hydrogen to space is, in most cases, independent
of photochemical details, being simply proportional to the
mixing ratio of water vapor at the level where molecular diffu-

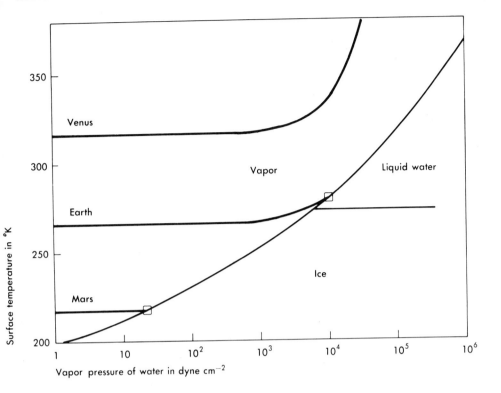

FIGURE 5. *The runaway greenhouse effect. The heavy curves show how surface temperatures increase, due to the greenhouse effect, as water vapor accumulates in the atmospheres of the inner planets. On Mars and Earth the increase is halted when the water vapor pressure is equal to the saturated vapor pressure (shown as the light line), and freezing or condensation occurs. Temperatures are higher on Venus because Venus is closer to the Sun and saturation is never achieved. The temperature runs away. (From Goody and Walker, 1972.)*

sion first becomes an important transport process. The rate of escape of hydrogen from the Earth is small today because the water vapor mixing ratio is small (a few times 10^{-6}) at heights above the tropopause cold trap. Since the primitive, wet atmosphere of Venus lacked a cold trap, mixing ratios of water vapor at upper levels would have been of the order of 1 and hydrogen escape rates could have been larger by many orders of magnitude than they are on Earth today (McElroy and Hunten, 1969; Smith and Gross, 1972; Hunten, 1973; Walker, 1975).

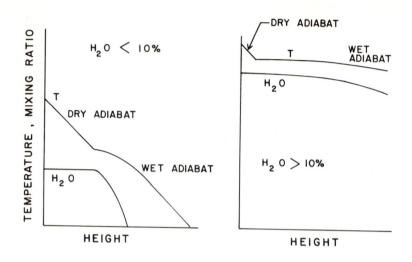

FIGURE 6. *Temperatures and water vapor mixing ratios as functions of height in convecting atmospheres containing little water (left) and abundant water (right). Both atmospheres are unsaturated near the ground; temperature decreases at the dry adiabatic lapse rate, and the water vapor mixing ratio is constant. Once condensation begins, release of latent heat decreases the temperature lapse rate. In the atmosphere with abundant water vapor the lapse rate is reduced very nearly to zero. Condensation reduces the water vapor mixing ratio to small values as height increases in the arid atmosphere, and the effect of latent heat on the temperature gradient becomes negligible. In the atmosphere with abundant water, on the other hand, both temperature and water vapor mixing ratio decrease very slowly with altitude.*

It appears that the rate of loss of hydrogen from Venus would have been limited not by the supply of hydrogen to the upper atmosphere (proportional to the water vapor mixing ratio), nor by the rate of photolysis of water vapor (about 10^{13} molecules cm^{-2} sec^{-1}), but by the flux of extreme ultraviolet radiation from the sun available to replace energy carried away from the upper atmosphere by the escaping hydrogen atoms (McElroy, 1974). Since a hydrogen atom escaping from Venus carries away 0.6 eV of energy and there are about 10^{12} eV cm^{-2} sec^{-1} in the extreme ultraviolet spectrum of the sun, a rough estimate of the rate of escape of hydrogen from the primitive atmosphere of Venus is 10^{12} atoms cm^{-2} sec^{-1} (Walker, 1977b). At this rate it would have taken only 600 million years to destroy a terrestrial ocean of water on Venus.

Even if the escape flux did not achieve this value it is unlikely that abundant water could have survived on Venus for much longer than a billion years. The oxygen left behind by escaping water was presumably consumed in reactions with the surface (Dayhoff *et al.*, 1967). The rate at which the surface consumed oxygen would have been limited by the rate of exposure to the atmosphere of fresh, unweathered rock (Walker, 1974). It is possible that the weathering rate was slow enough to permit oxygen to accumulate in the atmosphere (Walker, 1975), but it is equally possible that the rate of supply of fresh rock during the era of scum tectonics was more than large enough to consume oxygen as fast as it was produced.

In summary, then, the evolution of the atmosphere of Venus diverged from that of the Earth because Venus is too close to the Sun to permit water to condense on the surface. The greenhouse effect produced surface temperatures too high to permit carbonate minerals to form, and so carbon dioxide, like water, remained in the atmosphere. Because it contained water vapor as an abundant constituent, the atmosphere lacked a cold trap to concentrate the water at low levels. As a result, the water was rapidly destroyed by photolysis followed by escape of hydrogen to space and reaction of oxygen with the surface. While water vapor was necessary to cause the greenhouse effect to run away in the first place, the infrared opacity of the atmosphere has remained high since the water was lost (Pollack, 1969). Evidently other volatile elements driven into the atmosphere by the high surface temperature have served, together with carbon dioxide, to maintain the opacity, probably by contributing to the unbroken cloud cover of Venus.

MARS

On Mars the special circumstance is photochemical escape. The gravitational field of Mars is small enough to allow the escape into space of oxygen, carbon, and nitrogen atoms produced by exothermic chemical reactions in the upper atmosphere (Brinkmann, 1971; McElroy, 1972). Photochemical reactions buffer the hydrogen content of the atmosphere at a level that causes the rate of Jeans escape of hydrogen atoms to be twice the rate of photochemical escape of oxygen atoms (McElroy and Donahue, 1972; Liu and Donahue, 1976). Escape of oxygen therefore corresponds to destruction of water. Carbon escape, which occurs at a much slower rate, corresponds to destruction of carbon dioxide.

The rates of loss of oxygen and carbon are proportional to the flux of ionizing radiation from the sun as long as carbon dioxide is the dominant constituent of the atmosphere. We can therefore estimate the loss of water and carbon dioxide over the age of the planet by assuming that the solar output has remained constant. The total loss of water has been approximately 10^6 times the amount of water presently in the atmosphere but only 3×10^{-4} times the amount in the terrestrial ocean, while the total loss of carbon dioxide has been approximately equal to the amount of carbon dioxide now in the atmosphere Evidently escape from the atmosphere of Mars can not have been responsible for the loss of Earth-like amounts of water and carbon dioxide. Perhaps Mars was formed with a deficiency of volatiles, but I think it more likely that water and carbon dioxide have returned to the solid phase of the planet in the manner described above.

Nitrogen on Mars may not be susceptible to reincorporation into the solid phase, however, although the formation at the surface of nitrate salts has been suggested. Nitrogen on Earth enters the solid phase as a constituent of organic matter, but I assume that biological processes are geochemically unimportant on Mars. According to initial reports of the findings of the Viking Mission (Owen and Biemann, 1976; Nier et al., 1976), the atmosphere of Mars contains only 3% nitrogen (about 0.2 mbar), while there should at one time have been about 200 mbar of nitrogen in the atmosphere if Mars originally had an Earth-like complement of volatiles (including nitrogen in the Earth's crust (Turekian and Clark, 1975) and scaling by the planetary mass).

Photochemical escape of nitrogen occurs at a rate proportional to the mixing ratio of nitrogen as long as nitrogen is a minor constituent of the atmosphere. The nitrogen content of the atmosphere therefore decays exponentially with time if there is no source. The time constant for this decay has been estimated to lie between 8×10^7 and 6×10^8 years (McElroy, 1972; Walker, 1977b). Within the uncertainties there seems to be no difficulty in attributing the deficiency of nitrogen on Mars to photochemical escape.

The argon content of the Martian atmosphere requires some comment (Owen, 1974, 1976; Levine and Riegler, 1974; Levine, 1976; Fanale, 1976). Viking measured an argon concentration of only 1.5% (Owen and Biemann, 1976; Nier et al., 1976; Clark et al., 1976), which means that the atmosphere of Mars is deficient in argon relative to the atmosphere of Earth. This deficiency can not be attributed to photochemical escape or to reincorporation of argon into the solid phase of the planet.[1] It can be attributed to the rapid progression of Mars through the early and active phase of tectonic evolution.

[1]It is possible that low surface temperatures may permit adsorption of argon on the surface of regolith material (C. Alexander, private communication).

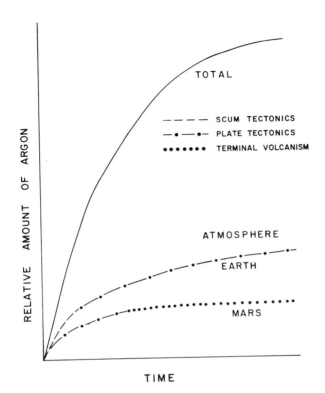

FIGURE 7. *Production of argon by the radioactive decay of potassium in Earth and Mars as a function of time, and the release of a portion of this argon to the atmosphere. Because of more rapid cooling of the interior, release slows down sooner on Mars than on Earth, leaving Mars with an atmosphere that is relatively deficient in argon.*

Argon is produced within the solid phase of a planet by the radioactive decay of potassium. The argon content of a planet therefore increases with the passage of time from an initially small value (see Fig. 7). Presumably most of the argon produced during the scum tectonics phase of planetary evolution is released to the atmosphere. Only some of the argon produced during the subsequent phase of plate tectonics is also released (90% of Earth's argon is still in the solid phase (Turekian, 1964; Fanale, 1976). Probably very little of the argon produced during the era of terminal volcanism is ever released. Therefore, Mars' atmosphere may be deficient in argon because little argon had been produced by the time the period of ready release of gas came to an end.

In summary, the evolution of the Martian atmosphere has
diverged from that of the terrestrial atmosphere because of the
low surface temperature caused by the planet's distance from
the Sun combined with its small size and small gravitational
field. Low surface temperature combined with small size has
permitted relatively rapid cooling of the interior. Low temperatures and temperature gradients within the planet have permitted water and carbon dioxide to return to the solid phase.
Low gravitational acceleration has permitted nitrogen to be
lost to space as a result of photochemical escape. Rapid
thermal evolution of the interior brought to an end the ready
release of argon to the atmosphere before much argon had been
formed.[2]

IMPLICATIONS FOR CHEMICAL EVOLUTION

It seems unlikely that conditions on Venus were ever favorable for the origin of life, although it is possible that
the atmospheric greenhouse ran away, as a result of gradually
increasing solar luminosity, some time after the origin of the
solar system (Pollack, 1971).

On Mars, on the other hand, I envisage a time when there
was an Earth-like amount of water, carbon dioxide, and nitrogen
at the surface. Presumably much of the water and carbon dioxide would have frozen out of the atmosphere, but the climate
may have been more equable when the atmosphere was not as tenuous as it is today (Sagan *et al.*, 1973; Gierasch and Toon,
1973). Besides, the combination of abundant ice and abundant

---

[2] *It has been reported by Owen and Biemann (1976) that ^{36}A
is 100 times less abundant in the atmosphere of Mars relative
to the mass of the planet than in the atmosphere of Earth.
This isotope is not produced by radioactive decay and may provide an indication of the total mass of volatiles released from
the interior of the planet. The implication is that Mars was
formed with a much lower volatile abundance than Earth, perhaps
only enough to produce 200 mbar of carbon dioxide, 300 mbar of
water, and 2 mbar of nitrogen (using terrestrial volatile ratios including the upper mantle, from Turekian and Clark (1975).
This deduction does not change my conclusion that low internal
temperatures have caused most of Mars volatiles to return to
the solid phase, but it means that the phrase "an Earth-like
complement of volatiles" must be interpreted generously. On
the other hand, it is not obvious that the ratio of inert gases
to condensable volatiles like carbon and hydrogen should be the
same on Mars as on Earth.*

volcanism during the era of scum tectonics may have been particularly favorable for chemical evolution (Miller and Orgel,
1974). Living conditions have deteriorated on Mars, as they
may eventually do on Earth. But the pace is faster on Mars.

ACKNOWLEDGMENT

The National Astronomy and Ionosphere Center is operated by
Cornell University under contract with the National Science
Foundation.

REFERENCES

Armstrong, R. L. (1968). A model for the evolution of strontium and lead isotopes in a dynamic earth, *Rev. Geophys. 6*,
175-199.
Armstrong, R. L., and S. M. Hein (1973). Computer simulation
of Pb and Sr isotope evolution of the Earth's crust and
upper mantle, *Geochim. Cosmochim. Acta 37*, 1-18.
Arrhenius, G., De, B. R., and Alfven, H. (1974). Origin of the
ocean, *in* "The Sea," Vol. 5 (E. Goldberg, ed.), pp. 839-
861, Wiley (Interscience), New York.
Brinkmann, R. T. (1971). Mars: Has nitrogen escaped? *Science
174*, 944-945.
Broecker, W. S. (1971). A kinetic model for the chemical composition of sea water, *Quat. Res. 1*, 188-207.
Cameron, A. G. W. (1973). Accumulation processes in the primi-
tive solar nebula, *Icarus 18*, 407-450.
Clark, B. C., Toulmin, P., Baird, A. K., Keil, K., and Rose,
H. J. (1976). Argon content of the Martian atmosphere at
the Viking I landing site: Analysis by X-ray fluorescence
spectroscopy, *Science 193*, 804-805.
Clark, S. P., Turekian, K. K., and Grossman, L. (1972). Model
for the early history of the earth, *in* The Nature of the
Solid Earth" (E. C. Robertson, ed.) pp. 3-18. McGraw-Hill,
New York.
Cruikshank, D. P., Morrison, D., and Lennon, K. (1973). Vol-
canic gases: Hydrogen burning at Kilauea Volcano, Hawaii,
Science 182, 277-279.
Dalgarno, A., and McElroy, M. B. (1970). Mars: Is nitrogen
present? *Science 170*, 167-168.
Dawson, J. B. (1971). Advances in kimberlite geology, *Earth
Sci. Rev. 7*, 187-214.
Dayhoff, M. O., Eck, R. V., Lippincott, E. R., and Sagan, C.
(1967). Venus: Atmospheric evolution, *Sience 155*, 556-558.

Dobrovolskis, A., and Ingersoll, A. P. (1975). Carbon-dioxide water clathrate as a reservoir of CO_2 on Mars, *Icarus 26,* 353-357.

Dzurisin, D., and Ingersoll, A. P. (1975). Seasonal buffering of atmospheric pressure on Mars, *Icarus 26,* 437-440.

Engel, A. E. J., Tison, S. P., Engel, C. G., Stickney, D. M., and Gray, E. J. (1974). Crustal evolution and global tectonics, *Bull Geol. Soc. Am. 85,* 843-858.

Fanale, F. P. (1971a). A case for catastrophic early degassing of the earth, *Chem. Geol. 8,* 79-105.

Fanale, F. P. (1971b). History of Martian volatiles: Implications for organic synthesis, *Icarus 15,* 279-303.

Fanale, F. P. (1976). Martian volatiles: Their degassing history and geochemical fate, *Icarus 28,* 179-202.

Fanale, F. P., and Cannon, W. A. (1971). Adsorption on the Martian regolith, *Nature 230,* 502-504.

Fanale, F. P., and Cannon, W. A. (1974). Exchange of adsorbed H_2O and CO_2 between the regolith and atmosphere of Mars caused by changes in surface insolation, *J. Geophys. Res. 79,* 3397-3402.

Flasar, F. M., and Goody, R. M. (1976). Diurnal behavior of water on Mars, *Planet. Space Sci. 24,* 161-181.

Gierasch, P. J., and Toon, O. B. (1973). Atmospheric pressure variations and the climate of Mars, *J. Atmos. Sci. 30,* 1502-1508.

Gold, T. (1964). Outgassing processes on the moon and Venus, *in* "The Origin and Evolution of Atmospheres and Oceans" (P. J. Brancazio and A. G. W. Cameron, eds), pp. 249-256. Wiley, New York.

Goody, R. M., and Walker, J. C. G. (1972). "Atmospheres." Prentice-Hall, Englewood Cliffs, New Jersey.

Green, H. W. (1972). A CO_2 charged asthenosphere, *Nature Phys. Sci. 238,* 2-5.

Grossman, L. (1972). Equilibrium condensation in the primitive solar nebula, *Geochim. Cosmochim. Acta 36,* 597-619.

Grossman, L., and Larimer, J. W. (1974). Early chemical history of the solar system, *Rev. Geophys. Space Phys. 12,* 71-101.

Hanks, T. C., and Anderson, D. L. (1969). The early thermal history of the earth, *Phys. Earth Planet. Inter. 2,* 19-29.

Hargraves, R. B. (1976). Precambrian geologic history, *Science 193,* 363-371.

Harris, A. W., and Kaula, W. M. (1975). A co-accretional model of satellite formation, *Icarus 25,* 516-524.

Hartmann, W. K. (1976). Planet formation: Compositional mixing and lunar compositional anomalies, *Icarus 27,* 553-559.

Hayes, J. M. (1967). Organic constituents of meteorites A review, *Geochim. Cosmochim. Acta 31,* 1395-1440.

Heald, E. F., Naughton, J., and Barnes, I. L. (1963). The chemistry of volcanic gases, use of equilibrium calculations in the interpretation of volcanic gas samples, *J. Geophys. Res. 68*, 545-557.

Hills, J. C. (1973). On the process of accretion in the formation of the planets and comets, *Icarus 18*, 505-522.

Holland, H. D. (1962). Model for the evolution of the Earth's atmosphere, *in* "Petrologic Studies: *A Volume in Honor of A. F. Buddington*" (A. E. J. Engel, H. L. James, and B. F. Leonard, eds.), pp. 447-477. Geological Society of America, New York.

Holland, H. D. (1964). On the chemical evolution of the terrestrial and cytherean atmospheres, *in* "The Origin and Evolution of Atmospheres and Oceans" (P. J. Brancazio and A. G. W. Cameron, eds), pp. 86-101. Wiley, New York.

Holland, H. D. (1976). The evolution of sea water, *in* "The Early History of the Earth" (B. F. Windley, ed), pp. 559-567. Wiley, London.

Houck, J. R., Pollack, J. B., Sagan, C., Schaack, D., and Decker, J. A. (1973). High-altitude spectroscopic evidence for bound water on Mars, *Icarus 18*, 470-480.

Huguenin, R. L. (1974). The formation of goethite and hydrated clay minerals on Mars, *J. Geophys. Res. 79*, 3895-3903.

Huguenin, R. L. (1976a). Surface oxidation: A major sink for water on Mars, *Science 192*, 138-139.

Huguenin, R. L. (1976b). Mars: Chemical weathering as a massive volatile sink, *Icarus 28*, 203-212.

Hunten, D. M. (1973). The escape of light gases from planetary atmospheres, *J. Atmos. Sci. 30*, 1481-1494.

Ingersoll, A. P. (1969). The runaway greenhouse: A history of water on Venus, *J. Atmos. Sci. 26*, 1191-1198.

Ingersoll, A. P. (1974). Mars: The case against permanent CO_2 frost caps, *J. Geophys. Res. 79*, 3403-3410.

Janssen, M. A., Hills, R. E., Thornton, D. D., and Welch, W. J. (1973). Venus: New microwave measurements show no atmospheric water vapor, *Science 179*, 994-997.

Johnston, D. H., McGetchin, T. R., and Toksöz, M. N. (1974). The thermal state and internal structure of Mars, *J. Geophys. Res. 79*, 3959-3971.

Kaula, W. M. (1975). The seven ages of a planet, *Icarus 26*, 1-15.

Kaula, W. M., and Bigeleisen, P. E. (1975). Early scattering by Jupiter and its collision effects in the terrestrial zone, *Icarus 25*, 18-33.

Kennedy, G. C., and Nordlie, B. E. (1968). The genesis of diamond deposits, *Econ. Geol. 63*, 495-503.

Kumar, S. (1976). Mercury's atmosphere: A perspective after Mariner 10, *Icarus 28*, 579-591.

Lambert, I. B., and Wyllie, P. J. (1970). Low velocity zone of the earth's mantle: Incipient melting caused by water, *Science 169*, 764-766.

Lambert, R. St. J. (1976). Archean thermal regimes, crustal and upper mantle temperatures, and a progressive evolutionary model for the Earth, *in* "The Early History of the Earth", (B. F. Windley, ed.), pp. 363-373. Wiley, London.

Leighton, R. B., and Murray, B. L. (1966). Behavior of carbon dioxide and other volatiles on Mars, *Science 153*, 136-144.

Levine, J. S. (1976). A new estimate of volatile outgassing on Mars, *Icarus 28*, 165-169.

Levine, J. S., and Riegler, G. R. (1974). Argon in the Martian atmosphere, *Geophys. Res. Lett. 1*, 285-287.

Lewis, J. S. (1972). Low temperature condensation from the solar nebula, *Icarus 16*, 241-252.

Lewis, J. S. (1974). The temperature gradient in the solar nebula, *Science 186*, 440-443.

Liu, S. C., and Donahue, T. M. (1976). The regulation of hydrogen and oxygen escape from Mars, *Icarus 28*, 231-246.

McElroy, M. B. (1972). Mars: An evolving atmosphere, *Science 175*, 443-445.

McElroy, M. B. (1974). Comment at conference on *The atmosphere of Venus*, Gooddard Institute for Space Studies, New York, October 15-17.

McElroy, M. B., and Donahue, T. M. (1972). Stability of the Martian atmosphere, *Science 177*, 986-988.

McElroy, M. B., and Hunten, D. M. (1969). The ratio of deuterium to hydrogen in the Venus atmosphere, *J. Geophys. Res. 74*, 1720-1739.

MacGregor, I. D., and Basu, A. R. (1974). Thermal structure of the lithosphere: A petrologic model, *Science 185*, 1007-1011.

Malin, M. C. (1974). Salt weathering on Mars, *J. Geophys. Res. 79*, 3888-3894.

Meadows, A. J. (1973). The origin and evolution of the atmospheres of the terrestrial planets, *Planet. Space Sci. 21*, 1467-1474.

Miller, S. L. and Orgel, L. E. (1974). "The Origins of Life on Earth." Prentice-Hall, Englewood Cliffs, New Jersey.

Mitchell, R. H., and Crocket, J. H. (1971). Diamond genesis - A synthesis of opposing views, *Mineral. Deposita 6*, 392-403.

Murray, B. C., and Malin, M. C. (1973). Polar volatiles of Mars - Theory versus observation, *Science 182*, 437-443.

Nier, A. O., Hanson, W. B., Seiff, A., McElroy, M. B., Spencer, N. W., Duckett, R. J., Knight, T. C. D., and Cook, W. S. (1976). Composition and structure of the Martian atmosphere: Preliminary results from Viking I, *Science 193*, 786-788.

Nordlie, B. E. (1968). Calculation of the basaltic gas phase composition, (*Abstr.*) *Geol. Soc. Am. Spec. Pap. 115*, 166.

Nordlie, B. E. (1972). Gases-Volcanic, *in* "The Encyclopedia of Geochemistry and Environmental Sciences" (R. W. Fairbridge, ed.), pp. 387-391. Van Nostrand-Reinhold, Princeton, New Jersey.

Orville, P. M. (1974). Crust-atmosphere interactions. Presented at conference on *The Atmosphere of Venus*, Goddard Institute for Space Studies, New York, October 15-17.

Owen, T. (1974). Martian climate: An empirical test of possible gross variations, *Science 183*, 763-764.

Owen, T. (1976). Volatile inventories on Mars, *Icarus 28*, 171-177.

Owen, T., and Biemann, K. (1976). Composition of the atmosphere at the surface of Mars: Detection of argon-36 and preliminary analysis, *Science 193*, 801-803.

Ozima, M. (1973). Was the evolution of the atmosphere continuous or catastrophic? *Nature Phys. Sci. 246*, 41-42.

Perry, E. C., and Tan, F. C. (1972). Significance of oxygen and carbon isotope determinations in early Precambrian cherts and carbonate rocks of southern Africa, *Geol. Soc. Am. Bull. 83*, 647-664.

Pollack, J. B. (1969). A nongray CO_2-H_2O greenhouse model of Venus, *Icarus 10*, 314-341.

Pollack, J. B. (1971). A nongray calculation of the runaway greenhouse: Implications for Venus' past and present, *Icarus 14*, 295-306.

Pollack, J. B., and Morrison, D. (1970). Venus: Determination of atmospheric parameters from the microwave spectrum, *Icarus 12*, 376-390.

Pollack, J. B., Pitman, D., Khare, B. N., and Sagan, C. (1970a). Goethite on Mars: A laboratory study of physically and chemically bound water in ferric oxides, *J. Geophys. Res. 75*, 7480-7490.

Pollack, J. B., Wilson, R. N., and Goles, G. G. (1970b). A reexamination of the stability of goethite on Mars, *J. Geophys. Res. 75*, 7491-7500.

Rama Murthy, V. (1976). Composition of the core and the early chemical history of the Earth, *in* "The Early History of the Earth" (B. F. Windley, ed.), pp. 21-31. Wiley, London.

Rasool, S. I., and DeBergh, C. (1970). The runaway greenhouse and the accumulation of CO_2 in the Venus atmosphere, *Nature 226*, 1037-1039.

Roedder, E. (1965). Liquid CO_2 inclusions in olivine bearing nodules and phenocrysts from basalts, *Am. Mineral. 50*, 1746-1782.

Ronov, A. B., and Yaroshevskiy, A. A. (1967). Chemical structure of the Earth's crust, *Geochemistry,* 1041-1066. (Transl. from *Geokhimiya).*

Ronov, A. B., and Yaroshevskiy, A. A. (1969). Chemical composition of the Earth's crust, *in* "The Earth's Crust and Upper Mantle" (P. J. Hart, ed.), pp. 37-57. Am. Geophys. Union Monograph 13, Washington, D. C.

Sagan, C. (1960). "The Radiation Balance of Venus." California Institute of Technology, Jet Propulsion Lab., Tech. Rept. No. 32-34.

Sagan, C. (1962). Structure of the lower atmosphere of Venus, *Icarus 1,* 151-169.

Sagan, C., Toon, O. B., and Gierasch, P. J. (1973). Climatic change on Mars, *Science 181,* 1045-1051.

Shaw, D. M. (1976). Development of the early continental crust. Part 2: Prearchean, Protoarchean and later eras, *in* "The Early History of the Earth" (B. F. Windley, ed.), pp. 33-53. Wiley, London.

Siever, R. (1974). Comparison of Earth and Mars as differentiated planets, *Icarus 22,* 312-324.

Smith, J. V. (1976). Development of the Earth-Moon system with implications for the geology of the early Earth, *in* The Early History of the Earth" (B. F. Windley, ed.), pp. 3-19. Wiley, London.

Smith, L. L., and Gross, S. H. (1972). The evolution of water vapor in the atmosphere of Venus, *J. Atmos. Sci. 29,* 173-178.

Sutton, J. (1976). Tectonic relationships in the Archaean, *in* "The Early History of the Earth" (B. F. Windley ed.), pp. 99-104. Wiley, London.

Sutton, J., and Watson, J. V. (1974). Tectonic evolution of continents in early Proterozoic times, *Nature 247,* 422-435.

Tozer, D. C. (1974). The internal evolution of planetary-sized objects, *The Moon 9,* 167-182.

Turekian, K. K. (1964). Degassing of argon and helium from the earth, *in* "The Origin and Evolution of Atmospheres and Oceans" (P. J. Brancazio and A. G. W. Cameron, eds.), pp. 74-85. Wiley, New York.

Turekian, K. K., and Clark, S. P. (1969). Inhomogeneous accumulation of the earth from the primitive solar nebula, *Earth Planet. Sci. Lett. 6,* 346-348.

Turekian, K. K., and Clark, S. P. (1975). The nonhomogeneous accumulation model for terrestrial planet formation and the consequences for the atmosphere of Venus, *J. Atmos. Sci. 32,* 1257-1261.

Veizer, J. (1976). $^{87}Sr/^{86}Sr$ evolution of seawater during geologic history and its significance as an index of crustal evolution, *in* "The Early History of the Earth" (B. F. Windley, ed.), pp. 569-578. Wiley, London.

Walker, J. C. G. (1974). Stability of atmospheric oxygen, *Am. J. Sci.* 274, 193-214.

Walker, J. C. G. (1975). Evolution of the atmosphere of Venus, *J. Atmos. Sci.* 32, 1248-1256.

Walker, J. C. G. (1976a). Formation of the inner planets, *Monthly Notes Astronomical Soc. Southern Africa* 35, 2-8.

Walker, J. C. G. (1976b). Implications for atmospheric evolution of the inhomogeneous accretion model of the origin of the earth, *in* "The Early History of the Earth." (B. F. Windley, ed.), pp. 337-346. Wiley, London.

Walker, J. C. G. (1977a). Origin of the atmosphere: History of the release of volatiles from the solid earth, *in* "Chemical Evolution of Early Precambrian" (C. Ponnamperuma, ed.), pp. 1-11. Academic Press, New York.

Walker, J. C. G. (1977b). "Evolution of the Atmosphere." Macmillan, New York.

Walker, J. C. G., Turekian, K. K. and Hunten, D. M. (1970). An estimate of the present-day deep-mantle degassing rate from data on the atmosphere of Venus, *J. Geophys. Res.* 75, 3558-3561.

Weidenschilling, S. J. (1974). A model for the accretion of the terrestrial planets, *Icarus* 22, 426-435.

Weidenschilling, S. J. (1976). Accretion of the terrestrial planets. II, *Icarus* 27, 161-170.

White, D. E., and Waring, G. A. (1963). Volcanic emanations, *U. S. Geol. Surv. Prof. Paper 440-K.*

8

THE EVOLUTION OF H_2O AND CO_2 ON EARTH AND MARS

Joel S. Levine

Atmospheric Environmental Sciences Division
NASA Langley Research Center, Hampton, Virginia

1. INTRODUCTION

In this chapter the evolutionary histories of water (H_2O) and carbon dioxide (CO_2) on Earth and Mars will be discussed. A number of pre-Viking papers have considered the evolution of H_2O and/or CO_2 on Mars (Sagan, 1967; Johnson, 1969; Fanale, 1971a, 1976; McElroy, 1972; Owen, 1974a,b, 1976; Levine and Riegler, 1974; Levine, 1976a; Huguenin, 1976). The evolution of H_2O and CO_2 on Venus has recently been discussed by Walker (1975). H_2O and CO_2 are by far the two most abundant volatiles outgassed from the Earth's interior (Rubey, 1951 and 1955). In addition both of these gases strongly absorb and then re-emit the infrared radiation emitted at the surface of Earth, Mars, and Venus and are the gases most responsible for the Greenhouse temperature enhancement on all three planets. Terrestrial atmospheric H_2O vapor and, to a smaller degree, CO_2 raise the mean radiative equilibrium temperature (determined by our distance from the Sun and global albedo) from about 240 to 288°K. CO_2 is responsible for increasing the radiative equilibrium temperature of Mars from about 220°K to the measured mean surface temperature of about 230°K. Hence, to a large extent atmospheric H_2O and CO_2 by virtue of the Greenhouse temperature enhancement control the climatology of the terrestrial planets.

It is apparent that H_2O and CO_2 have had most divergent evolutionary histories on Earth and Mars. The surface of Earth is covered by several kilometers of liquid water with atmospheric CO_2 only a trace species (330 ppm), whereas the surface of

Mars is devoid of liquid water and the atmosphere, which is predominantly CO_2 (∿95%), contains only trace amounts of H_2O vapor
(∿300 ppm).

   Although this chapter deals primarily with the evolutionary histories of H_2O and CO_2 on Earth and Mars, the evolution
of the other species in the terrestrial atmosphere is briefly
discussed in Section 2. Understanding the natural evolution of
the terrestrial atmosphere is particularly relevant right now,
due to the current interest and concern about the effects of
anthropogenic activities on the troposphere, stratosphere, and
climate. Section 3 deals with evolutionary histories of H_2O and
CO_2 on Mars in a scenario to suggest that Mars may have outgassed
as much as 10-100 times more H_2O and 2-20 times more CO_2 over
its history than previously believed. The possible fate of
these huge quantities of H_2O and CO_2 on Mars is described in
Section 4. The likely reason for the divergent evolutionary
paths of H_2O and CO_2 on Earth and Mars is discussed in Section
5.

2. THE EVOLUTION OF THE EARTH'S ATMOSPHERE

   According to current thinking, the Earth's atmosphere and
oceans accumulated as a result of volatile outgassing from the
Earth's interior (Rubey, 1951, 1955). The accumulation of the
Earth's atmosphere and oceans could have been gradual and continuous over geological time (Rubey, 1951, 1955) or could have
been rapid and early, and has essentially come to a stop (Fanale, 1971b). The bulk of the outgassed volatiles such as H_2O,
CO_2, molecular nitrogen, neon, ^{36}Ar, krypton, and xenon that
formed the present atmosphere and oceans were originally
trapped, either chemically bound and/or physically adsorbed,
within the solid Earth during the planetary accretion process,
while other volatiles such as ^{4}He and ^{40}Ar resulted from radiogenic decay of uranium and thorium, and ^{40}K, respectively, within the solid Earth. The Earth's atmosphere is extremely deficient in the primordial or nonradiogenic noble gases (neon, ^{36}Ar,
krypton, and xenon) compared to their cosmic abundance, as
pointed out by Brown (1949) and Suess (1949) and summarized in
Table I. This deficiency of the nonradiogenic noble gases has
been interpreted as evidence to suggest that the Earth's present atmosphere is not a remnant of the primordial solar nebula,
but, rather is secondary, having outgassed from the Earth's interior. The primordial noble gases are believed to have been
selectively excluded during the planetary accretion process,
since they do not solidify at likely accretion temperatures
and do not form chemical compounds. The chemical inertness of
argon, krypton, and xenon precludes their loss of identity, and

TABLE I. *Deficiency of Primordial Noble Gases in the Earth's Atmosphere*[a]

Element	Terrestrial to Cosmic Abundance
Neon	$\sim 10^{-10}$
Argon-36	$\sim 10^{-8}$
Krypton	$\sim 10^{-6}$
Xenon	$\sim 10^{-6}$

[a]*From Berkner and Marshall (1965), based on Brown (1949) and Suess (1949).*

their large molecular mass precludes their atmospheric escape, suggesting that these gases may be used as indices of volatile outgassing on the terrestrial planets.

To complete the present atmospheric inventory we must include biologically produced species such as oxygen, nitrous oxide, methane, and ammonia and photochemically-produced species such as atomic oxygen, hydroxyl, and atomic hydrogen resulting from the photolysis of outgassed H_2O, ozone, and the oxides of nitrogen, hydrogen, and chlorine. We must understand the natural variability, including the evolution of the terrestrial atmosphere before we can accurately assess and critically evaluate the effect of anthropogenic activities on the atmosphere such as the effects of increased carbon dioxide on global climate or the possible inadvertant depletion of atmospheric ozone due to (1) nitrogen oxides produced by supersonic transports, (2) increased nitrous oxide production resulting from increased worldwide use of nitrogen fertilizer, and (3) chlorine released photochemically from the chlorofluoromethanes used as propellants in aerosol spray cans (Levine, 1976b).

With the exception of the noble elements, the gases trapped in the solid Earth during the planetary accretion process probably reflected the cosmic abundance of the primordial solar nebula that gave birth to the Earth and the solar system about 4.5 billion years ago. The cosmic abundance of elements is given in Table II (Cameron, 1973). The conditions within the primordial solar nebula must have been highly reducing due to the overwhelming presence of excess hydrogen. It has been suggested that in the early history of the Earth, before geological differentiation (before the migration of metallic iron to the core) the outgassed volatiles were probably highly reduced, consisting of methane and smaller amounts of ammonia, molecular

TABLE II. *Cosmic Abundance of Selected Elements*[a]

Element	Cosmic abundance
H	2.60×10^{10}
He	2.10×10^{9}
O	2.36×10^{7}
C	1.35×10^{7}
N	2.44×10^{6}
Ne	2.36×10^{6}
^{36}Ar	2.28×10^{5}
Kr	64.40
Xe	7.10

[a]*Compilation of abundances normalized to Si $= 10^{6}$ (Cameron, 1973).*

hydrogen, and H_2O (Holland, 1962, 1964). This early methane-ammonia atmosphere appears to be the most suitable environment for initiating the development of complex biological molecules and the chemical evolution of life (Oparin, 1938; Miller, 1953; Miller and Urey, 1959; Urey, 1959; Ponnamperuma and Gabel, 1968). It is believed that the present composition of Jupiter's atmosphere is similar to that of the Earth's early atmosphere, which gave rise to complex biological molecules, the precursors of life. The presence and possible methods of remote detection of these organic molecules on Jupiter have been recently dis-cussed by Levine and Rogowski (1975).

The migration of metallic iron to the Earth's core had the effect of increasing the average degree of oxidation of the mantle and crust material. Hence, after core formation, the outgassed volatiles would have been much less chemically re-duced: methane was replaced by CO_2, ammonia was replaced by molecular nitrogen, and molecular hydrogen was replaced by H_2O (Holland, 1962, 1964). The time of the formation of the Earth's core is uncertain, with best estimates suggesting core formation at least 3.5 billion years ago, during the Earth's first billion years (Johnson, 1969). The duration of an early methane-ammon-ia atmosphere is another unresolved question with estimates ranging from 10^{5} to 10^{8} years (Rubey, 1955) to as long as 10^{9} years (McGovern, 1969).

Estimates of the total inventory of outgassed H_2O, CO_2, and argon over the history of the Earth are given in Table III. The most abundant gas released from the Earth's interior, H_2O (4.3×10^5 gm cm^{-2}), condensed out of the atmosphere and is present mainly in liquid form in the Earth's extensive oceans. Over the Earth's history some 150 gm cm^{-2} of outgassed H_2O has been photodissociated, with the subsequent thermal escape of about 20 gm cm^{-2} of atomic hydrogen resulting in an accumulation of about 130 gm cm^{-2} of molecular oxygen (Johnson, 1969).

The bulk of the CO_2, the next most abundant outgassed volatile (7.9×10^4 gm cm^{-2}), precipitated out of the atmosphere in the presence of liquid water in the form of calcium carbonate ($CaCO_3$) in the ocean sediments. The outgassed nitrogen, the third most abundant outgassed volatile, being quite chemically inert, has largely remained in the atmosphere and forms the bulk of the Earth's present atmosphere (78.08%). About 10% of atmospheric nitrogen has been removed from the atmosphere and placed in geological deposits, and a small fraction (about 10^{-8}) of atmospheric nitrogen is fixed each year, mainly by biological processes. Most of the fixed nitrogen is eventually returned to the atmosphere within a few years or decades (Johnson, 1969).

Radiogenic volatiles (^{40}Ar and ^4He) constitute an important class of outgassed species in the Earth's atmosphere. On the cosmic abundance scale, the isotopic abundance ratio of ^{36}Ar, ^{38}Ar, and ^{40}Ar is 84.2, 15.8, and 0%, respectively (Cameron, 1973), while the measured isotopic abundance ratio of argon in the terrestrial atmosphere is 0.33, 0.06, and 99.6%, respectively (Banks and Kockarts, 1973). Therefore, the bulk of the argon in the terrestrial atmosphere has resulted from the radioactive decay of ^{40}K, as opposed to being trapped from the primordial solar nebula during the planetary accretion process (as were ^{36}Ar and ^{38}Ar). Unlike ^4He, argon is a relatively heavy atom and cannot escape from the atmosphere, thus ^{40}Ar has accumulated over geological time to become the third most abundant permanent constituent of the terrestrial atmosphere (0.934x%). The role radiogenic outgassing and the evolution of ^{40}Ar and ^4He in the Martian atmosphere have been discussed (Levine and Riegler, 1974; Levine et al., 1974).

From a geochemical point of view, the presence of large amounts of free oxygen (20.95%) in the terrestrial atmosphere is puzzling. The solid Earth is underoxidized and outgassed volatiles do not contain free oxygen. In fact, the interaction with surface material and volcanic gases is actually a significant sink for free oxygen. Crustal rock weathering and the continual exposure of fresh minerals that are underoxidized are therefore significant drains on the atmospheric oxygen supply.

TABLE III. Estimates of Terrestrial Inventory of Outgassed H_2O, CO_2, and Argon

	Total Inventory ($\times 10^{20}$ gm)	Normalized with Respect to ^{40}Ar (gm cm^{-2})	Reference
H_2O	16,600	2.52×10^4	Rubey (1951, 1955)
	18,000-27,000	$2.73 - 4.09 \times 10^4$	Poldervaart (1955)
	18,000	2.73×10^4	Vinogradov (1967)
	22,000	3.33×10^{4}[a]	Ronov and Yaroshevsky (1969)
CO_2	910	1.38×10^3	Rubey (1951, 1955)
	2,550	3.86×10^3	Poldervaart (1955)
	4,000	6.06×10^3 [a]	Ronov and Yaroshevsky (1969)
^{40}Ar	0.66	1.00[b]	Banks and Kockarts (1973)
^{36}Ar	0.002	3×10^{-3}	Banks and Kockarts (1973)

[a]Values adopted in this chapter.

[b]The Earth's atmosphere contains 13 gm cm^{-2} of ^{40}Ar.

The chronology of the rise of free oxygen in the Earth's atmosphere is still an unresolved problem. The evolution of oxygen has been investigated by Berkner and Marshall (1965), Brinkmann (1969), and more recently by Walker (1974, 1976) and Margulis et al., (1976).
While the exact chronology for the rise of free oxygen is not known, we can place limitations on the levels of oxygen in the Earth's early atmosphere. Analysis of the oldest rocks (about 3 billion years old) and their associated minerals and mineral systems can be used to place limitations on the levels of oxygen and other trace gases in the atmosphere beneath which they were deposited (Holland, 1962; Abelson, 1966; Cloud, 1968). Detrital uranium and pyrite at various localities indicate that before 1.8 to 2.0 billion years ago, the atmosphere contained very little, if any, free oxygen (Cloud, 1968).
In their investigation, Berkner and Marshall concluded that over its history, the bulk of the Earth's oxygen evolved from the biological process of photosynthesis - as opposed to the

abiotic process - the release of oxygen due to the photolysis of water vapor with the accompanying exoshperic escape of atomic hydrogen. The Berkner and Marshall chronology for the rise of atmospheric evolution is summarized in Table IV.

An interesting aspect of the Berkner and Marshall oxygen chronology is the prediction of greater than present levels of oxygen during the Carboniferous period. There is also the suggestion that oxygen concentrations fluctuated in a dampened oscillation around the present level before actually reaching the present level.

The photodissociation of water vapor and the evolution of oxygen in the Earth's atmosphere has been reexamined by Brinkman (1969). Brinkman challenged Berkner and Marshall's conclusion that over most of geologic time, the atmospheric oxygen level has been $<10^{-3}$ the present atmospheric level. Brinkmann's calculations indicate atmospheric oxygen levels of greater than 0.25 P.A.L. over 99% of the Earth's history. This estimate is a factor of 250 times higher than the upper limit of Berkner and Marshall and implies that our atmosphere could have been highly oxidized over a large fraction of geologic time even in the absence of the widespread photosynthetic activity operating today. As Brinkmann points out, his calculation suggesting the presence of 0.25 P.A.L. of oxygen over 99% of the Earth's history is directly contrary to the geologic record and chemical evolution implications, which strongly suggest that earlier than 2 billion years ago the Earth's atmosphere was anaerobic.

TABLE IV. The Growth of Oxygen in the Atmosphere[a]

	Oxygen level (P.A.L.)
Prior to 2.7 b.y. ago	10^{-4}
Prior to 600 m.y. ago (Precambrian)	10^{-3}
600 m.y. ago (Cambrian)	10^{-2}
420 m.y. ago (Late Silurian)	10^{-1}
280 m.y. ago (Carboniferous)	3 (?)
Present	1

[a]Oxygen levels in present atmospheric level (P.A.L.) (Berkner and Marshall, 1965).

The reformation of water vapor as a possible sink for atmospheric oxygen was underestimated in Brinkmann's analysis. Brinkmann used an atomic hydrogen escape flux of 1.6×10^9 cm^{-2} sec^{-1}, which is a factor of 10 higher than the currently accepted value of 1×10^8 cm^{-2} sec^{-1}. Brinkmann's dissociation rate for H_2O of 1.6×10^9 cm^{-2} sec^{-1} and the accepted atomic hydrogen escape of 1×10^8 cm^{-2} sec^{-1} means that only about 6% of the atomic hydrogen atoms escape from the atmosphere, with 94% eventually oxidized back to water vapor, leaving very little free atmospheric oxygen.

The following discussion is largely based on material contained in Margulis *et al.* (1976). They argue that it is very unlikely that the photolysis of water vapor could have produced an aerobic atmosphere before the origin of oxygen-producing photosynthetic organisms (Walker, 1974). They add that the abiotic oxygen levels calculated by Berkner and Marshall and Brinkmann "are almost certainly too high,...," since "both studies erred by assuming that every photolysis is followed by escape."

Margulis *et al.* (1976) also point out that Berkner and Marshall and Brinkmann" also erred by neglecting oxygen consumption in reactions with reduced volcanic gases, principally hydrogen." At present, the outgassing of hydrogen from volcanoes appears to be comparable to the exospheric escape of atomic hydrogen (Margulis, *et al.*, 1976). In the early Precambrian stage, Margulis *et al.* (1976) suggest that the volcanic source of hydrogen was larger than the abiotic source of oxygen, which resulted in low atmospheric oxygen levels.

The appearance of blue-green algae between 2 and 3 billion years ago introduced a new and important source of free atmospheric oxygen. The transition from a hydrogen-rich reducing atmosphere to an oxidizing atmosphere has been estimated as occurring at about 2.0 ± 0.2 billion years ago by Cloud (1968). This atmospheric transition coincides with the time that banded iron formations were replaced with red beds in the sedimentary record. The deposition of banded iron formations required an anaerobic atmosphere (Cloud, 1968), while the deposition of the red beds required an aerobic atmosphere (Van Houten, 1973).

In summary, while the exact chronology for the accumulation of oxygen in the Earth's atmosphere is not known, there is considerable evidence to suggest that the early atmosphere contained little or no free oxygen. At some time in the Earth's history oxygen rose to, and perhaps even exceeded, the present levels.

Closely related to the evolution of free oxygen in the Earth's atmosphere is the origin and evolution of atmospheric ozone (Berkner and Marshall, 1965; Ratner and Walker, 1972; Levine, 1976b). The photochemistry of atmospheric ozone has recently received considerable attention due to the possibility of inadvertent depletion of ozone by the following anthropogenic activities: (1) nitrogen oxides produced by supersonic transports, (2) increased nitrous oxide production by soil bacteria resulting from increased worldwide use of nitrogen fertilizer, and (3) chlorine released photochemically from the chlorofluoromethanes used as propellants in aerosol spray cans (Stratospheric Ozone Depletion, 1975).

Other minor atmospheric species that have an important role in atmospheric chemistry and atmospheric evolution include several biologically produced gases such as nitrous oxide produced at the surface by soil bacteria and methane produced at the surface from the decomposition of organic matter in swamps and paddy fields. Nitrous oxide is the main natural source of odd nitrogen compounds that catalytically destroy stratospheric ozone Methane is an important photochemical source of atmospheric carbon dioxide and hydrogen atoms.

As shown in this section, the composition of the Earth's atmosphere can be quantitatively understood in terms of the outgassing of trapped and radiogenic volatiles, photochemistry, and biologically produced species.

3. THE EVOLUTION OF H_2O and CO_2 ON MARS

Our classic picture of Mars is that of a dry, desertlike planet completely devoid of surface liquid water with only minute traces of atmospheric water vapor (several hundred ppm) that is highly variable with latitude and season (Barker, 1976). The Martian atmosphere is predominantly CO_2. Ground-based spectroscopic measurements indicate a CO_2 partial pressure of 5.2 ± 0.3 mbar, corresponding to about $14.0 \pm .80$ gm cm^{-2} of CO_2 (Young, 1969).

Over the years various investigators have estimated the total outgassed H_2O and CO_2 on Mars. The total amount of H_2O outgassed on Mars over its history has been estimated on the basis of the current thermal escape of atomic hydrogen, a photolysis product of H_2O. Using this approach various investigators have estimated that Mars outgassed between 1 and 3 x 10^2 gm cm^{-2} of H_2O (Sagan, 1967; Johnson, 1969; McElroy, 1972). These estimates suggest that Mars outgassed several orders of magnitude less H_2O than the Earth.

Another approach to estimate the total amount of H_2O outgassed over a planet's history is based on using ^{40}Ar as an index of outgassed H_2O (Kulp, 1951). However, ^{40}Ar on Mars cannot be measured from the Earth since its spectroscopic resonance lines lie in the inaccessible extreme ultraviolet (at

1066 and 1048 $\overset{o}{A}$. Using an Earth-analogous outgassing model
that assumed Mars was geologically differentiated and that the
^{40}K abundance of the Martian interior was comparable to that
of the terrestrial interior per unit mass, Levine and Riegler
(1974) suggested that over its history, ^{40}Ar may have accumulated to as much as 28% of the Martian atmosphere. This
quantity of argon was consistent with a brief announcement in
Izvestia that the Soviet Mars-6 lander indirectly inferred the
presence of "several tens of percent of some inert gas ...most
probably argon" in the Martian atmosphere (Moroz, 1974). A
recent detailed account of the Mars-6 measurements concludes by
noting that "the inert gas concentration in the Martian atmosphere amounts to 30 ± 10% by volume. The experiment performed
is the first direct evidence for the presence of an inert gas,
evidently argon, as an important component of the Martian atmosphere" (Istomin and Grechnev, 1976).

The amount of argon in the Martian atmosphere reported by
the Mars-6 investigators was consistent with the amount
of argon predicted using an Earth-analogous outgassing model
(Levine and Riegler, 1974). This leads to the suggestion that
the Earth's total volatile outgassing ratio, given in Table II,
may be applicable to Mars and that argon could in fact be used
as an index of total outgassed H_2O, as first suggested by Kulp
(1951), as well as an index of other outgassed volatiles such
as CO_2 and nitrogen.

To consider the implications of argon on the evolutionary
history of the Martian atmosphere a special symposium entitled
"The Stability and Evolution of the Martian Atmosphere" was
held at the 1975 annual meeting of the American Geophysical
Union (*EOS, Trans. Am. Geophys. Union 56*, 460, 1975). Most of
the papers presented at this symposium were published in a
special pre-Viking "argon on Mars" issue of *Icarus* (Vol. 28,
June 1976). Under the assumptions that (1) ^{40}Ar is an index of
volatile outgassing, (2) the terrestrial volatile outgassing
ratios given in Table III are applicable to Mars, and (3) the
Martian atmosphere contains several tens of percent of ^{40}Ar, it
was estimated that Mars may have outgassed as much as 10^5 gm
cm^{-2} of H_2O over its history (Levine, 1975, 1976a; Fanale, 1975,
1976) and as much as 10^4 gm cm^{-2} of CO_2 over its history (Levine
and Riegler, 1974; Levine, 1975, 1976a; Owen, 1975, 1976;
Fanale, 1975, 1976).

However, the concentration of ^{40}Ar in the Martian atmosphere
was sharply revised downward in the summer of 1976 with the
successful orbital insertion and subsequent landing of the two
Viking spacecraft. Three different Viking I instruments, the
entry mass spectrometer, the lander mass spectrometer, and the
X-ray fluorescence experiment, all reported ^{40}Ar concentrations
1-2% (Nier *et al.*, 1976; Owen and Biemann, 1976; Clark

et al., 1976a). In addition, the lander mass spectrometer measured the atmospheric ^{40}Ar: ^{36}Ar ratio to be 2750 ± 500 (Owen and Biemann, 1976), an order of magnitude greater than the ^{40}Ar:^{36}Ar ratio in the terrestrial atmosphere of about 330 (Banks and Kockarts, 1973). These results were confirmed by the same three instruments on Viking II.

The pre-Viking prediction of some 28% of ^{40}Ar in the Martian atmosphere was based on the assumption of comparable ^{39}K and ^{40}K abundances per unit mass on both Earth and Mars (Levine and Riegler, 1974). It now appears that Mars may be significantly deficient in ^{39}K and ^{40}K, at least in its soil composition. The Viking I and II lander X-ray fluorescence spectrometers conducted elemental analyses of the Martian regolith soil at the two widely separated landing sites (Chryse Planita and Utopia Planitia) and reported remarkably similar results. At both sites, the regolith soil contains abundant amounts of silicon and iron with significant concentrations of magnesium, aluminum, sulfur, calcium, and titanium. However, potassium was not detected in the analysis at either site, suggesting an abundance of <0.25 wt% (Clark, *et al.*, 1976b). The average potassium content in the Earth's crust is 1.25 to 2.1% (Heier and Billings, 1970). Hence, the Viking measurements suggest that the average crustal material of Mars contains at least 5-8 times less potassium than the Earth (Clark *et al.*, 1976b). The measured ^{40}Ar concentration in the Martian atmosphere is consistent with the lack of detectable potassium in the Martian soil.

There is some discussion in the literature as to whether ^{36}Ar or ^{40}Ar is the better index of volatile outgassing. Atmospheric ^{36}Ar was originally trapped from the primordial solar nebula during the planetary accretion process. However, it is widely believed that ^{36}Ar and the other primordial noble gases were selectively excluded during the accretion process. Hence, ^{36}Ar may not be a good index of the trapping of the other more chemically active primordial volatiles. On the other hand, ^{40}Ar, the radiogenic decay product of ^{40}K, is not a primordial volatile, but is related to both the concentration of ^{40}K (and hence ^{39}K in the solid planet) and the extent of volatile outgassing. The use of ^{40}Ar as an index of total potassium is most useful, since relationships between the bulk potassium content of solid solar system bodies and trapped chemically active volatiles have been established (Lewis, 1973; Fanale, 1976). Since Viking instruments measured both the atmospheric ^{40}Ar concentration as well as the atmospheric ^{40}Ar:^{36}Ar ratio, we shall use both ^{36}Ar and ^{40}Ar as indices to estimate the lower and upper limits, respectively, of both H_2O and CO_2 outgassing on Mars.

Assuming the validity of the terrestrial H_2O:^{40}Ar ratio to
Mars, the presence of 1.5% ^{40}Ar in the Martian atmosphere suggests that Mars outgassed some 7.0 x 10^3 gm cm^{-2} of H_2O. The
terrestrial H_2O:^{36}Ar ratio (1.11 x 10^7:1 in gm cm^{-2}) suggests
that Mars outgassed a factor of ten less H_2O than estimated
using the ^{40}Ar abundance or about 7.0 x 10^2 gm cm^{-2} of H_2O.
However, Lewis (1973) has suggested that the bulk H_2O content
of Mars may be enriched by as much as six times that of Earth
per unit mass due to the complete retention of tremolite (hydrous calcium silicate) during the accretion of Mars. The retention of tremolite on Mars would result in 4.2 x 10^3 to 4.2
x 10^4 gm cm^{-2} of H_2O based on Viking-measured concentrations of
^{36}Ar and ^{40}Ar, respectively.

The CO_2 estimate of Ronov and Yaroshevsky (1969) is a factor
4.4 times greater than that of Rubey (1951 and 1955) and yields
a terrestrial CO_2:^{36}Ar ratio of 2.0 x 10^6:1 in gm cm^{-2} and a
CO_2:^{40}Ar ratio of 6.1 x 10^3:1 in gm cm^{-2}. These outgassing
ratios coupled with the Viking ^{36}Ar and ^{40}Ar measurements suggest that Mars may have outgassed as much as 10^2-10^3 gm cm^{-2}
of CO_2 over its history. These estimates are considerably
greater than the 30-80 gm cm^{-2} of outgassed CO_2 previously suggested (Johnson, 1969; Murray and Malin, 1973). However, the
values for outgassed CO_2 estimated here are not inconsistent
with a recent estimate of 1.5 x 10^3 gm cm^{-2} of outgassed CO_2
suggested by Clark and Mullin (1976) from their analysis of
Martian glaciation and the flow of solid CO_2.

The amounts of H_2O and CO_2 that Mars may have outgassed estimated here, 4.2 x 10^3-10^4 gm cm^{-2} of H_2O (corresponding to
1.5 - 15.0 bar of H_2O) and 10^2-10^3 gm cm^{-2} of CO_2 (corresponding
to 37-370 mbar of CO_2) suggest that Mars outgassed considerably
more volatiles than previously believed. The analysis of the
Viking-measured nitrogen isotopes in the Martian atmosphere
suggests that Mars may have outgassed as much as 30 mbar of
nitrogen over its history (McElroy et al., 1976), again suggesting that Martian outgassing was considerably more extensive than
previously believed.

4.   WHERE ARE THE VOLATILES?

To continue with this scenario of the evolution of H_2O and
CO_2 on Mars, we must consider the fate of the large quantities
of outgassed H_2O and CO_2 implied by the Viking argon measurements and the applicability of the terrestrial outgassing to
Mars. For many years it was assumed that exospheric escape
was the major volatile sink on Mars. Photochemical calculations
indicate that over its history Mars could have lost only about

1-3×10^2 gm cm^{-2} of H_2O (Sagan, 1967: Johnson, 1969; McElroy, 1972) and only about 1 gm cm^{-2} of CO_2 (McElroy, 1972) via exospheric escape. Hence, it appears that exospheric escape is not an efficient way to lose large amounts of H_2O and CO_2 on Mars.

It now appears that large amounts of outgassed H_2O and CO_2 may be stored in the Martian regolith. As much as 10^3 gm cm^{-2} of H_2O may be physically adsorbed to the regolith (Fanale and Cannon, 1974) and as much as 4×10^4 gm cm^{-2} of H_2O may be chemically bound in geothite and clay (Fanale, 1976). In addition, ground ice in the form of global ice lenses could accommodate up to 5×10^4 gm cm^{-2} of H_2O (Fanale, 1976). Photostimulated oxidation or "chemical weathering" could have irreversibly removed 10^2-10^5 gm cm^{-2} of H_2O from the Martian atmosphere over its history (Huguenin, 1976). Recent Viking orbiter measurements suggest that the northern polar cap remnant may consist of water ice rather than CO_2 ice as had previously been believed. This suggests that the northern polar cap remnant may be an important H_2O sink.

The regolith may also be an important sink for outgassed CO_2. As much as 7×10^3 gm cm^{-2} of CO_2 may be stored in the regolith as physically adsorbed CO_2 and chemically bound in the form of carbonates (Fanale, 1976). Photostimulated oxidation or "chemical weathering" could have irreversibly removed 10^1-10^4 gm cm^{-2} of CO_2 from the Martian atmosphere over its history (Huguenin, 1976). The total amount of CO_2 stored in the northern polar cap remnant is at present unclear due to recent Viking orbiter measurements that suggest the presence of water ice rather than CO_2 ice. In the pre-Viking period, Sagan (1971) suggested that the complete vaporization of the northern polar cap remnant would release more than 10^3 gm cm^{-2} of CO_2 into the atmosphere over the whole planet. While the composition of the northern polar cap remnant is unclear at the present time it appears that regolith storage and "chemical weathering" are potential massive sinks for the large amounts of outgassed H_2O and CO_2 estimated in this chapter.

5. DIVERGENT EVOLUTIONARY PATHS OF H_2O AND CO_2 ON EARTH AND MARS

The divergent evolutionary paths of H_2O and CO_2 on Earth and Mars can be simply explained by each planet's distance from the Sun. In the absence of an atmosphere in the very early history of Earth and Mars, the planet-Sun distance determined the radiative equilibrium surface temperature of the planet: about 220^0K for Mars and about 250^0K for Earth (Rasool and DeBergh, 1970). The planet-Sun distance not only initially

determined the planet's surface temperature, but also determined the tropospheric temperature of each planet. This temperature in turn controls the H_2O vapor capacity of the troposphere, since the saturation vapor pressure is strongly temperature dependent, i.e., the higher the tropospheric temperature the more H_2O vapor can be accommodated. Furthermore, the
surface temperature determines the state (liquid or solid) of
the condensed H_2O vapor after it reaches the atmospheric saturation vapor pressure.

Rasool and DeBergh (1970) showed that the surface and troposphere temperatures of Mars were so low that outgassed H_2O
vapor could not accumulate in the atmosphere for long before
the partial pressure of the atmospheric H_2O vapor reached the
saturation vapor pressure of ice. Since the partial pressure
of H_2O vapor cannot exceed the saturation vapor pressure, any
additional H_2O vapor released to the atmosphere condensed out
of the atmosphere onto the surface in the form of water frost
and remained in that state. Hence, the surface temperature
ceased to rise via the Greenhouse temperature enhancement for
any additionally outgassed H_2O vapor, once the saturation vapor
pressure of ice was reached.

Earth, being closer to the Sun, had a somewhat higher radiative equilibrium surface temperature than Mars. After H_2O
vapor outgassing began on Earth, the partial pressure of H_2O
vapor approached the saturation vapor pressure and H_2O vapor
then condensed out of the atmosphere in liquid form. The
Earth's primordial surface temperature allowed the condensed
vapor to remain in the liquid state to accumulate to form
oceans at the Earth's surface. The greater tropospheric temperature on the Earth compared to Mars permitted larger amounts
of H_2O vapor to accumulate before the saturation vapor pressure was reached, and condensation began. The greater build-
up of H_2O vapor in the Earth's atmosphere compared to that of
Mars increased the surface temperature of Earth by some $40^{\circ}K$
due to the Greenhouse effect (Rasool and DeBergh, 1970). The
outgassing of additional H_2O vapor after the saturation vapor
pressure was reached did not result in an additional Greenhouse
temperature enhancement on either Earth or Mars.

The presence of liquid water oceans controlled the evolutionary path of the outgassed CO_2 on Earth by removing the
bulk of the CO_2 from the atmosphere by the formation of calcium
carbonate ($CaCO_3$) presently found in the ocean sediments,
whereas the absence of liquid water on the surface of Mars
allowed the outgassed CO_2 to remain in the atmosphere, where it
is the overwhelming constituent ($\sim95\%$) of the Martian atmosphere.

The nature of Earth and Mars, two dissimilar planets, with different atmospheric compositions and surface conditions, can be explained by the same theory of atmospheric evolution as outlined in the scenario presented here. We have increased our understanding of our planet's atmospheric composition, photochemistry, climatology, dynamics, and circulation from studies of the other planets in the solar system. For example, the analytical techniques involving coupled diffusion and photochemical equations used to study the photochemistry and possible inadvertent depletion of stratospheric ozone in the 1970s were, by and large, developed by planetary scientists in the 1960s to better understand the composition and stability of the CO_2 atmospheres of Mars and Venus. Comparative studies of the other planets, and their atmospheres, are important in obtaining a better understanding of the past, present, and future of our planet.

REFERENCES

Abelson, P. H. (1966). Chemical events on the primitive Earth, *Proc. U. S. Natl. Acad. Sci., 55*, 1365-1372.

Banks, P. M., and Kockarts, G. (1973). "Aeronomy," Part A, pp. 18-25. Academic Press, New York.

Barker, E. S. (1976). Martian atmospheric water vapor observations: 1972-74 apparition, *Icarus, 28,* 247-268.

Berkner, L. V. and Marshall, L. C. (1965). On the origin and rise of oxygen concentrations in the Earth's atmosphere, *J. Atmos. Sci. 22,* 225-261.

Brinkmann, R. T. (1969). Dissociation of water vapor and evolution of oxygen in the terrestrial atmosphere, *J. Geophys. Res. 74,* 5355-5368.

Brown, H. (1949). Rare gases and the formation of the Earth's atmosphere, *in* "The Atmospheres of the Earth and Planets" (G. P. Kuiper, ed), pp. 260-268. Univ. of Chicago Press, Chicago, Illinois.

Cameron, A. G. W. (1973). Abundances of the elements in the solar system, *Space Sci. Rev. 15,* 121.

Clark, B. R., and Mullin, R. P. (1976). Martian glaciation and the flow of solid CO_2, *Icarus 27,* 215-228.

Clark, B. C., Toulmin, P., Baird, A. K., and Rose, H. J. (1976a). Argon content of the Martian atmosphere at the Viking I landing site: Analysis by X-ray fluoresence spectroscopy, *Science 193,* 804-805.

Clark, B. C., Baird, A. K., Rose, H. J., Toulmin, P., Keil, K., Castro, A. J., Kelliher, W. C., Rowe, C. D., and Evans, P. H. (1976b). Inorganic analyses of Martian surface samples at the Viking landing sites, *Science 194*, 1283-1288.

Cloud, P. E. (1968). Atmospheric and hydrospheric evolution on the primitive Earth, *Science, 160*, 729-736.

Fanale, F. P. (1971a). History of Martian volatiles: Implications for organic synthesis, *Icarus 15*, 279-303.

Fanale, F. P. (1971b). A case for catastrophic early degassing of the Earth, *Chem. Geol. 8*, 79-105.

Fanale, F. P. (1975). Regolith storage of volatiles and an Earth-analogous Mars degassing model, EOS, *Trans Am. Geophys. Union 56*, 406.

Fanale, F. P. (1976). Martian volatiles: Their degassing history and geochemical fate, *Icarus 28*, 179-202.

Fanale, F. P., and Cannon, W. A. (1974). Exchange of absorbed H_2O and CO_2 between the regolith and atmosphere of Mars caused by changes in surface insolation, *J. Geophys. Res. 79*, 3397-3402.

Heier, K. S., and Billings, G. K. (1970). *In* "Handbook of Geochemistry" (K. H. Wedpohl, ed), Vol. 2, Part 2. Springer-Verlag, New York.

Holland, H. D. (1962). Models for the evolution of the Earth's atmosphere. Petrologic studies: A volume to honor A. F. Buddington, *Geol. Soc. Am.* 447-477.

Holland, H. D. (1964). On the chemical evolution of the terrestrial and cytherean atmospheres *In* "The Origin and Evolution of Atmospheres and Oceans" (P. J. Brancazio and A. G. W. Cameron, eds.), pp. 86-101. Wiley, New York.

Huguenin, R. L. (1976). Mars: Chemical weathering as a massive volatile sink, *Icarus 28*, 203-212.

Istomin, V. G., and Grechnev, K. V. (1976). Argon in the Martian atmosphere: Evidence from the Mars-6 descent module, *Icarus 28*, 155-158.

Johnson, F. S. (1969). Origin of planetary atmospheres, *Space Sci. Rev. 9*, 303-324.

Kulp, J. L. (1951). Origin of the hydrosphere, *Bull. Geol. Soc. Am. 62*, 326-329.

Levine, J. S. (1975). Argon on Mars!: Where is the water? *EOS, Trans. Am. Geophys. Union 56*, 405.

Levine, J. S. (1976a). A new estimate of volatile outgassing on Mars, *Icarus 28*, 165-169.

Levine, J. S. (1976b). The making of the atmosphere, *Advan. Eng. Sci.*, NASA CP-2001, Vol. 3, pp. 1191-1201.

Levine, J. S., Keating, G. M., and Prior, E. J. (1974). Helium in the Martian atmosphere: Thermal loss considerations, *Planet Space Sci. 22*, 500-503.

Levine, J. S., and Riegler, G. R. (1974). Argon in the Martian atmosphere, *Geophys. Res. Lett. 1*, 285-287.

Levine, J. S., and Rogowski, R. S. (1975). Fluorescence detection of organic molecules in the Jovian atmosphere, *Origins of Life 6*, 395-399.

Lewis, J. S. (1973). The origin of the planets and satellites, *Tech. Rev. 76*, 21-35.

Margulis, L., Walker, J. C. G., and Rambler, M. (1976). Reassessment of roles of oxygen and ultraviolet light in Precambrian evolution, *Nature, 264*, 620-624.

McElroy, M. B. (1972). Mars: An evolving atmosphere, *Science 175*, 443-445.

McElroy, M. B., Yung, Y. L., and Nier, A. O. (1976). Isotropic composition of nitrogen: Implications for the past history of Mars' atmosphere, *Science 194*, 70-72.

McGovern, W. E. (1969). The primitive Earth: Thermal models of the upper atmosphere for a methane-dominated environment, *J. Atmos. Sci. 26*, 623-635.

Miller, S.L. (1953). Production of amino acids under possible primitive Earth conditions, *Science 117*, 528-529.

Miller, S. L., and Urey, H. C. (1959). Organic compound synthesis on the primitive Earth, *Science 130*, 245-251.

Moroz, V. I., (1974). Portrait of Planet Mars, *Izvestia*, March 1974.

Murray, B. C., and Malin, M. C. (1973). Polar volatiles on Mars - Theory vs. observation, *Science 182*, 437-443.

Nier, A. O., Hanson, W. B., Seiff, A., McElroy, M. B., Spencer, N. W., Duckett, R. J., Knight, T. C. D., and Cook, W. S. (1976). Composition and structure of the Martian atmosphere: Preliminary results from Viking I, *Science 193*, 786-788.

Oparin, A. I. (1938). "Origin of Life." MacMillan, New York.

Owen, T. (1974a). What else is present in the Martian atmosphere? *Comments Astrophys. Space Phys. 5*, 175-180.

Owen, T. (1974b). Martian climate: An empirical test of possible gross variations, *Science 183*, 763-764.

Owen, T. (1975). The permanent composition of the Martian atmosphere, *EOS, Trans. Am. Geophys. Union 56*, 405.

Owen, T. (1976). Volatile inventories on Mars, *Icarus 28*, 171-177.

Owen, T., and Biemann, K. (1976). Composition of the atmosphere at the surface of Mars: Detection of argon-36 and preliminary analysis, *Science 193*, 801-803.

Poldervaart, A. (1955). Chemistry of the Earth's crust, *Geol. Soc. Am., Spec. Paper 62*, pp. 119-144.

Ponnamperuma, C., and Gabel, N. W. (1968). Current status of chemical studies on the origin of life, *Space Life Sci. 1*, 64-96.

Rasool, S. I., and DeBergh, C. (1970). The runaway greenhouse and the accumulation of CO_2 in the Venus atmosphere, *Nature 226*, 1037-1039.

Ratner, M. I., and Walker, J. C. G. (1972). Atmospheric ozone and the history of life, *J. Atmos. Sci. 29*, 803-808.

Ranov, A. B., and Yaroshevsky, A. A. (1969). Chemical composition of the Earth's crust. The Earth's crust and upper mantle, *Geophys. Monograph. 13*, Am. *Geophys. Union*, pp. 37-57.

Rubey, W. W. (1951). Geologic history of sea water: An attempt to state the problem, *Geol. Soc. Am.*, *Bull. 62*, pp. 1111-1147.

Rubey, W. W. (1955). Development of the hydrosphere and atmosphere with reference to probable composition of the early atmosphere, *Geol. Soc. Am. Spec. Paper 62*, pp. 631-650.

Sagan, C. (1967). Origins of atmospheres of Earth and planets, *In* "International Dictionary of Geophysics," pp. 97-104. Pergamon, Oxford.

Sagan, C. (1971). The long-winter model of Martian biology. A speculation, *Icarus 15*, 511-514.

Stratospheric Ozone Depletion: Hearings before the Subcommittee on the Upper Atmosphere of the Committee on Aeronautical and Space Sciences of the United States Senate, Ninety-fourth Congress, First Session, September 8, 9, 15 and 17, 1975. Part I. U. S. Government Printing Office, 1975, 554 pages.

Suess, H. E. (1949). Die Haufigkeit der Edelgase auf der Erde und in Kosmos, *J. Geol. 57*, 600-607.

Urey, H. C. (1959). Primitive planetary atmospheres and the origin of life, *in* "The Origin of Life on Earth," Vol I, *(Symp. Intern. Union Biochem.)* MacMillan, New York. pp. 16-22.

Van Houten, F. B. (1973). Origins of red beds: A review, 1961-1972, *Ann. Rev. Earth Planet. Sci. 1*, 39-61.

Vinogradov, A. P. (1967). The formation of the ocean, *Izv. Akad. Nauk SSSR. Ser. Geol. (4)*, 3-9.

Walker, J. C. G. (1974). Stability of atmospheric oxygen. *Am. J. Sci. 274*, 193-214.

Walker, J. C. G. (1976). Implications for atmospheric evolution of the inhomogeneous accretion model of the origin of the Earth, *in* "The Early History of the Earth" pp. 537-546. (B. F. Windley, Ed.), Wiley, New York.

Walker, J. C. G. (1975). Evolution of the atmosphere of Venus, *J. Atmos. Sci. 32*, 1248-1256.

Young, L. D. G. (1969). Interpretation of high-resolution spectra of Mars I. CO_2 abundance and surface pressure derived from the curve of growth, *Icarus 11*, 386-389.

9

THE ATMOSPHERE OF MARS: DETECTION OF KRYPTON AND XENON

T. Owen

Department of Earth and Space Sciences
State University of New York, Stony Brook, New York

K. Biemann, J. E. Biller

Department of Chemistry, Massachusetts Institute of Technology,
Cambridge, Massachusetts

D. W. Howarth

Guidance and Control Systems
Litton Industries, Woodland Hills, California

A. L. LaFleur

Department of Chemistry, Massachusetts Institute of Technology
Cambridge, Massachusetts

Krypton and xenon have been discovered in the martian atmosphere with the mass spectrometer on the second Viking lander. Krypton is more abundant than xenon. The relative abundances of the krypton isotopes appear normal, but the ratio of xenon-129 to xenon-132 is enhanced on Mars relative to the terrestrial value for this ratio. Some possible implications of these findings are discussed.

We have previously reported the detection of ^{36}Ar and the establishment of upper limits on Ne, Kr, and Xe in the atmosphere of Mars, using the mass spectrometer on the first Viking lander (Owen and Biemann, 1976; Biemann *et al.*, 1976a). The upper limit on krypton was close to the value that would be predicted if the ^{36}Ar/Kr ratio on Mars were identical to that of Earth. It thus seemed important to try to increase the sensitivity of the experiment with the hope of actually detecting this important element.

The successful deployment of the second Viking lander (VL2) on Mars afforded us the opportunity to conduct this experiment. We had established from tests of the two instruments during the cruise from Earth to Mars that the background in the mass spectrometer on VL2 was much lower than that in the instrument on the first lander, making it a superior instrument for the detection of trace amounts of atmospheric gases. It was also clear from experience gained with the operation of VL1 that the best way to maintain this low instrumental background would be to perform the atmospheric analyses prior to any analyses of the martian soil. A delay in obtaining the first soil sample for our instrument with VL2 provided the opportunity to design an optimized enrichment sequence for detecting trace gases in the atmosphere.

The basic procedure has already been described (Owen and Biemann, 1976; Biemann *et al.*, 1976a). We use a chemical scrubber to remove over 99% of the CO and CO_2 in a given sample of martian atmosphere, admit a new sample, again remove CO_2, and repeat this procedure a preselected number of times to build up the concentration of trace constituents. The sample is exposed to a drying agent during each cycle to remove water generated by the CO_2 scrubber. The sample is then admitted to the mass spectrometer through a molecular leak for analyses. Using sequences of five and ten cycles, we obtained enrichments of 4 and 6.3 over the yield from a single cycle. Mass spectra of the sample enriched 6.3 times showed an indication of krypton, but the identification was not conclusive. After evaluating the performance of the instrument, we changed the internal timing of the sequence and obtained a tenfold enrichment with 15 cycles. This sample was analyzed 16 times by the mass spectrometer with the electron multiplier gain increased by a factor 5.3 over its nominal value. These spectra gave clear evidence of the presence of krypton. The analysis was subsequently repeated after an additional 15 cycles with a multiplier gain of 28 times nominal.

The results are shown in Fig. 1, which indicates the appearance of the averaged mass spectrum in the vicinity of the krypton and xenon isotopes. The characteristic isotopic pattern of krypton is clearly evident; the broad peak at $m/e = 80$ is an

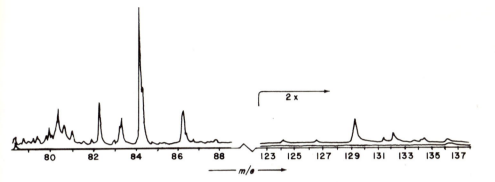

FIGURE 1. Mass spectra of enriched samples of the martian atmosphere in the region of krypton and xenon. The spectra are averages of nine scans, the lower lines are averages of three background scans. The vertical scale is linear; it has been increased by a factor of 2 for xenon.

artifact caused by the high partial pressure of argon in the instrument. In the case of xenon, ^{129}Xe is much more abundant relative to the other isotopes on Mars compared with the distribution in terrestrial atmospheric xenon. The absence of organic compounds of high molecular weight in the soil (Biemann, et al., 1976b) the absence of spectroscopically active compounds of high molecular weight in the atmosphere (Owen, 1974; Barth, 1974), and the clean instrumental background in this mass range make us quite confident that the peaks shown in this region of Fig. 1 are chiefly caused by xenon on Mars.

We plan to obtain additional data with both VL1 and VL2 ultimately using a more extreme enrichment sequence, which should lead to an additional gain in the concentration of trace gases by a factor between 1.5 and 2.0. Such a sequence involves some risk to the instrument, however, since the partial pressure of argon then becomes dangerously high for the capabilities of the ion pump. We will not, therefore, attempt this procedure until the postconjunction period, near the end of the extended mission.

It is not yet possible to compute accurate abundances for these two gases, or even to give precise values for the ratios of their isotopes. At this low level of detection, the instrumental response is noisy, nonlinear, and distorted by memory effects (degassing from the pump). There is, however, no question that ^{129}Xe is much more abundant than ^{132}Xe and ^{131}Xe, rather than almost equal to the other two isotopes as is the case on Earth. The mass spectrometer must be recalibrated in a manner that duplicates experimental conditions on Mars. Such tests are now being conducted with the third gas chromatograph-mass spectrometer (GCMS) that was built for this mission.

Meanwhile, we can point out some important consequences of this discovery, based on the available data. First, we can support our previous assertion that it is very unlikely that Mars had a massive, original atmosphere (produced, for example, by accretional heating) which subsequently was reduced to its present state by extensive solar wind sweeping. If that were the case, we would expect the ratio of ^{36}Ar to total krypton to be much lower than the value found in the earth's atmosphere or in the primordial gas in meteorites, since the argon would be swept off more efficiently than krypton in the upper atmosphere where diffusive separation occurs. The low total abundance of ^{36}Ar now in the martian atmosphere therefore implies the following possibilities; (1) that Mars was very depleted in volatiles at the time of its formation--an unlikely condition in view of its distance from the sun, (2) that a large fraction of the primitive atmosphere was swept away in a manner that led to no mass discrimination, or (3) that the planet has simply not out-gassed as much as Earth--the conclusion we favor (Owen and Biemann, 1976; Biemann *et al.*, 1976a).[1]

The second consequence of this observation stems from there being, evidently, more krypton than xenon in the martian atmosphere. This result is important because the reverse is true in the primordial gas component of ordinary or carbonaceous chondrites (Signer, 1964; Mazor *et al.*, 1970). The abundance pattern and the absolute abundances of noble gases in this primordial component closely match the abundances of the noble gases observed in the earth's atmosphere except for xenon (8), a fact that led to the prediction that this same pattern should exist in the martian atmosphere (Owen, 1974; Fanale, 1971). This prediction was based on the assumption that the fractionation process that produced the primordial gas component in the meteorites must have occurred prior to the formation of the planets and thus must have affected the noble gas distribution in the accreting planetary material in the same way. Most meteorites

---

[1]*One could argue that the noble gases were originally present on Mars in a "cosmic" ratio, in which case the $^{36}Ar/Kr$ ratio would be 2.5 x 10^3 and Kr/Xe 9 (Cameron, 1973). The Ar/Kr ratio would then have been reduced to its present earthlike value coincidentally by solar wind sweeping. This seems to ask a lot both from coincidence and from the solar wind, but a definitive test would be the detection of neon, which may yet be possible. (The enhancement of ^{129}Xe also argues against a strictly cosmic noble gas inventory.)*

come from the region of the asteroid belt, Mars is between
Earth and the asteroids, hence it seemed reasonable that Mars
should show the same type of noble gas abundance pattern.

However, in the earth's atmosphere, xenon is deficient compared with the primordial gas in meteorites, and this is exactly the situation we are now finding on Mars (Fig. 2). The
xenon deficiency on Earth has been attributed to the preferential adsorption of xenon in shales and other sedimentary material after it was outgassed (Canales et al., 1968; Fanale and
Cannon, 1971). One is thus led to the tentative conclusion that
similar processes have been active on Mars; perhaps in association with the epochs of fluvial erosion that have left their
imprint on the planet's surface. An alternative (or supple-
mentary) suggestion is that some of the xenon could be adsorbed
in the regolith (F. P. Fanale, private communication).

At this stage of our investigation of martian xenon, we
can only be sure of the enhancement of ^{129}Xe. A determination
of the relative abundances of the other isotopes requires
additional analyses of enriched samples. It is generally
agreed that ^{129}Xe anomalies in meteoritic and terrestrial gas
samples result from the production of this isotope by decay of
extinct ^{129}I (Reynolds, 1963; Pepin, 1964). We find that the
ratio of ^{129}Xe to ^{132}Xe is 2.5 (+2 or -1); the terrestrial atmospheric value is 0.97, and in the carbonaceous and ordinary
chondrites, values as high as 4.5 and 9.6 have been reported

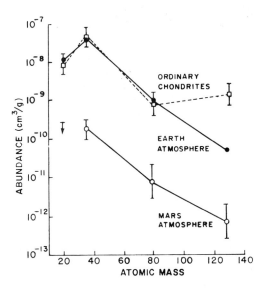

FIGURE 2. The abundances of noble gases in ordinary chondrites and the atmospheres of Earth and Mars. Error bars are
only approximate since systematic errors remain to be assessed.

(Signer, 1964; Mazor *et al.*, 1970; Reynolds, 1963; Pepin, 1964). There is a tendency for more ^{129}Xe in meteorites of types C-3 and C-4 than types C-1 or C-2, which may be attributable to the greater ability of the coarse-grained C-3 and C-4 types to retain the ^{129}Xe than the fine-grained C-1 and C-2 types, according to Mazor *et al.* (1970). One might therefore conjecture that a planetary atmosphere derived from a partially degassed veneer enriched in material of type C-1 would show an enhancement in an atmosphere that had suffered massive losses at an appropriate interval after formation of the planet.

These are only two of several hypotheses that could explain the observations, however, and we shall study these problems in more detail after we have obtained results from the experiments planned for the end of the extended mission.

ACKNOWLEDGMENTS

We thank A. Tomassian and W. Dencker for optimizing the mass spectrometer and atmospheric filter assembly, respectively. We are indebted to A. V. Diaz, E. M. Ruiz, and R. Williams for their assistance. This work was supported by research contracts NAS 1-10493 and NAS 1-9684 from the National Aeronautics and Space Administration.

REFERENCES

Barth, C. A. (1974). *Annu. Rev. Earth Planet. Sci. 2*, 33.

Biemann, K., Owen, T., Rushneck, D. R., and LaFleur, A. L. (1976a), *Science 194*, 76.

Biemann, K., Oro, J., Toulmin, P. III, Orgel, L. E., Nier, A. O., Anderson, D. M., Simmonds, P. G., Flory, D., Diaz, A. V., Rushneck, D. R., and Biller, J. E. (1976b). *Science 194*, 72.

Cameron, A. G. W. (1973). *Space Sci. Rev. 15*, 121. Canalas, R. A., Alexander, E. C., and Manuel, O. K. (1968). *J. Geophys. Res. 73*, 3331.

Fanale, F. (1971). *Icarus 15*, 279.

Fanale, F. R., and Cannon, W. A. (1971). *Earth Planet. Sci. Lett. 11*, 362.

Mazor, E., Heymann, D., and Anders, E. (1970). *Geochim. Cosmochim. Acta 34*, 781.

Owen, T. (1974). *Comments Astrophys. Space Phys. 5*, 175.

Owen, T., and Biemann, K. (1976). *Science 193*, 801.

Pepin, R. O. (1964). *In* "The Origin and Evolution of Atmospheres and Oceans" (P. J. Brancazio and A. G. W. Cameron, eds.), Chap. 9. Wiley, New York.
Reynolds, J. H. (1963). *J. Geophys. Res. 68*, 2939.
Signer, P. (1964). *In* "The Origin and Evolution of Atmospheres and Oceans" (P. J. Brancazio and A. G. W. Cameron, eds.), Chap. 8. Wiley, New York.
Wasson, J. (1969). *Nature (London 223*, 163.

10

COMPOSITION OF THE ATMOSPHERE
AT THE SURFACE OF MARS:
DETECTION OF ARGON-36
AND PRELIMINARY ANALYSIS

Tobias Owen

Department of Earth and Space Sciences
State University of New York, Stony Brook, New York

K. Biemann

Department of Chemistry, Massachusetts Institute of Technology
Cambridge, Massachusetts

The composition of the martian atmosphere was determined by the mass spectrometer in the molecular analysis experiment. The presence of argon and nitrogen was confirmed and a value of 1 to 2750 ± 500 for the ratio of argon-36 to argon-40 was established. A preliminary interpretation of these results suggests that Mars had a slightly more massive atmosphere in the past, but that much less total outgassing has occurred on Mars than on Earth.

The objective of the Viking molecular analysis experiment is twofold: to detect and identify the organic compounds, if any, present in the surface of Mars, and to determine periodically the composition of the lower atmosphere (Anderson *et al.*, 1972). The central part of the instrumentation for this experiment is a mass spectrometer, coupled to a gas chromatograph for the organic analysis and, by way of a molecular leak, to a gas sample reservoir. Although the instrument was designed primarily for the detection of organic compounds in the gas chromatographic mode (Biemann, 1974), the mass spectrometer's high sen-

sitivity (dynamic range, six to seven orders of magnitude), high
mass range (m/e 12-200), and resolution (1:200 at m/e 200; better at lower mass) were used to advantage in determining the
composition of the atmosphere, particularly its minor constituents. The penalty one pays for resolution and sensitivity is
a certain loss of accuracy, mainly because the residual background in the instrument becomes more significant and the long-
term reproducibility of the fragmentation pattern is lowered.

Since the more important questions concerning the composition of the martian atmosphere centered around the minor components and certain isotopic ratios, an attempt was made to
optimize the experiment toward that goal. In particular, the
detection of even traces of N_2 was deemed to be extremely important because previous data (Barth et al., 1969; Dalgarno and
McElroy, 1970) suggested that it must be a minor component or
could be almost completely absent (McElroy, 1972). One of the
major problems in a mass spectrometric determination of N_2 in
the martian environment is the interference of CO^+ (from CO or
CO_2) with the ion current of N_2^+ at m/e 28. For this reason the
gas reservoir of the instrument is coupled via separate valves
to two cavities, one containing Ag_2O and LiOH, for the oxidation
of CO to CO_2 and absorption of all CO_2, and the other containing
$Mg(ClO_4)_2$ for the removal of the resulting water. During the
fourth and fifth day after the landing of Viking 1 (July 20,
1976) a total of six atmospheric analyses were performed at
approximately 6-hour intervals. In the first four of these analyses, CO and CO_2 were removed; in the last two analyses, samples of unaltered atmosphere were used. During the third analysis the spectrometer shut down temporarily, leaving us with a
total of five sets of mass spectral scans. Analysis of these
spectra gave the averaged results shown in Table I.

It is clear from these results that the N_2 content is consistent with earlier limits (Barth et al., 1969; Dalgarno and
McElroy, 1970) and corroborates the results of the upper atmosphere analysis (Nier et al., 1976) performed during the descent
of the Viking 1 lander. The argon content is much lower than
most recently suggested (Istomin and Grechnev, 1976) and again
supports the value found for the upper atmosphere (Nier et al.,
1976)

Of most importance is that the high sensitivity of our mass
spectrometer permitted the determination of the abundance of
[36]Ar which was found to be about ten times lower than the value
corresponding to the terrestrial ratio of [36]Ar to [40]Ar. Neon,
krypton, and xenon could not be detected at the limits shown
in Table II. It should be pointed out that for this preliminary
analysis the quoted detection limit of neon is higher than the
amount of [36]Ar actually determined. This is because of the low
instrument background and little interference at m/e 36 (the

TABLE I. *Preliminary data on the abundances of gases de-tected in the martian atmosphere*

Component[a]	Abundance (%)
Carbon dioxide	95
Oxygen	0.1 to 0.4
Nitrogen	2 to 3
Argon	1 to 2
$^{36}Ar/^{40}Ar$ ratio	1:2750 ± 500

[a]*Variable amounts of water, 0.16 percent carbon monoxide, and 0.03 ppm of ozone have been found in the martian atmosphere by ground-based or spacecraft observations Owen (1974a,b). Because of surface adsorption the mass spectrometric data on H_2O are rather meaningless. The reasons for the CO remaining undetermined are mentioned in the text, and ozone is far below our detection limit.*

TABLE II. *Upper limits on the gases not detected in the atmosphere*

Gas	Initial upper limit (ppm)[a]
Neon	10
Krypton	20
Xenon	50

[a]*We expect to decrease these limits for the analysis of the atmosphere under conditions of enrichment of the minor constituents.*

absence of a corresponding signal at m/e 35 excludes any contribution from HCl) and because of the finite contributions of $^{40}Ar^{2+}$ to m/e 20 even at 45 ev.

Low concentrations (less than a few percent) of CO cannot be detected with our system, because of the interference by the large amount of CO_2 in the two analyses of the concomitant removal of CO when removing the CO_2 in the analyses where it was exposed to Ag_2O. Also, the determination of O_2 is reliable only in the unaltered atmosphere, because of the possibility

that some of the Ag_2O produces O_2. For this reason, the values
of O_2 in Table I are from only two measurements.

The determination of other minor constituents, including
the isotope ratios for N_2, must await further refinement of the
data and additional analyses, which are planned. The abundances
of ^{13}C and [18]O as determined from the m/e 44, 45, and 46 signals
appear to be equal to the terrestrial values within the accuracy
of our preliminary measurements.

At this stage of the analysis and acquisition of data, it
would be premature to draw any firm conclusions regarding the
outgassing history of the planet. We can, however, relate some
of our results to a theoretical consideration of this problem
using ideas developed in anticipation of the experiment (Owen,
1974a,b).

The noble gases provide a particularly useful measure of
the degree of planetary outgassing. The report by Istomin and
Grechnev (1976) that the martian atmosphere might contain as
much as 35 percent [40]Ar was widely interpreted to indicate that
Mars had outgassed as thoroughly as Earth, and led to the prediction that massive amounts of concurrently produced volatiles
had either escaped from the planet or were presently buried in
the regolith (Levine, 1976; Owen, 1976; Fanale, 1976; Hugenin,
1976). The discovery that the [40]Ar abundance is only 1 - 2% of
the present atmosphere indicates that the other volatiles should
be proportionately reduced in such models. But the fact that
the [36]Ar/[40]Ar ratio is less on Mars than it is on Earth by a
factor of 10 suggests that the interpretation of martian outgassing may not be quite so straightforward. The [40]Ar may be
anomalously abundant on Mars. It should be safer to scale the
abundances of other volatiles relative to [36]Ar, the nonradiogenic isotope.

If we compare the absolute abundances of [36]Ar on Mars and
Earth, taking ratios of the mass of gas to the mass of the re-
spective planet, we find that the amount in the martian atmosphere is approximately 100 times smaller than the terrestrial value. This implies that the total degassing of Mars is
less complete than that of Earth by about the same factor, if
we ignore the possibility of [36]Ar escape. The ratios $CO_2/[36]Ar$
and $N_2/[36]Ar$ are both roughly ten times smaller in the martian
atmosphere than in Earth's inventory of volatiles. It is especially interesting that the relative abundances of N_2 and CO_2
now in the martian *atmosphere* are very similar to the values
found in Earth's *inventory*.

One interpretation of these results is that Earth and Mars
have a similar bulk composition, so gases are produced in similar proportions but degassing and subsequent weathering on Mars
have been much less complete. The present martian atmosphere
would then represent about one-tenth the mass of the total out-

gassed volatiles exclusive of water. The corresponding amount of water is equivalent to a layer a few tens of meters deep.

The missing volatiles may be trapped in subsurface permafrost (H_2O) and at the polar caps (H_2O, CO_2), chemically bound in the soil (nitrates, oxides, carbonates), and some portion must have escaped from the planet (McElroy, 1972). In this view, the ^{40}Ar abundance represents an anomaly, being roughly ten times more abundant than predicted. This discrepancy could result from several causes, for example, an enrichment of potassium in the martian crust relative to Earth, or a different degassing history for ^{40}Ar compared with the other volatiles, an interpretation that would be consistent with its radiogenesis from ^{40}K.

This Earth-analog model implies that the martian atmosphere was never much more than ten times as massive as it is now, producing a maximum surface pressure of $\sim$100 mbar. But with the possibility that CO_2 can be trapped at the poles and that large amounts of water could be present in the form of permafrost, we leave open the opportunity for cyclical or at least episodic variations of the mean climatic conditions on the planet, which would permit the formation of the sinuous channels by water erosion during temperate periods.[1]

The relative importance of the various processes for disposing of the missing volatiles, as well as an improved estimate of their total bulk, must await further analysis. It is reassuring to realize that the Viking landers have the capability for performing some of the most critical experiments needed to answer these questions.

ACKNOWLEDGMENTS

We thank our fellow team members, D. Anderson, L. Orgel, J. Oro, and P. Toulmin, III, and especially A. O. Nier, for helpful discussions and support. We thank J. E. Biller, A. V. Diaz, D. W. Howarth. A. L. LaFleur, E. M. Ruiz, D. R. Rushneck, and R. Williams for their innovative assistance in the development of this experiment. There are others, too numerous to mention, who have made essential comtributions; we thank them all. Supported by NASA research contracts NAS 1-10493 and NAS 1-9684.

[1]*Many investigators suggested this possibility in the wake of the Mariner 9 discovery of the channels. For a review of fluvial processes, see Masursky (1973), and Milton (1973). For a cyclical model of climate history, see Sagan et al. (1973).*

REFERENCES

Anderson, D. M., *et al.* (1972). *Icarus 16*, 111.

Barth, C. A., Fastie, W. G., Hord, C. W., Pearce, J. B.,
    Kelley, K. K., Stewart, A. I., Thomas, G. E., Anderson, G.
    P., and Raper, O. F. (1969). *Science 65*, 1004.

Stewart, A. I., Thomas, G. E., Anderson, G. P., and Raper, O.
    F. (1969). *Science 65*, 1004.

Biemann, K. (1974). *Origins Life 5*, 417.

Dalgarno, A., and MCElroy, M. B. (1970). *Science 170*, 167.

Istomin, V. G., and Grechniv, K. V. (1976). *Icarus 28*, 155.

Levine, J. S. (1976). *Icarus 28*, 165.

Fanale, F. P. (1976). *Icarus 28*, 179.

McElroy, M. B. (1972). *Science 175*, 443.

Nier, A. O., Hanson, W. B., Seiff, A., McElroy, M. B., Spencer,
    M. W., Duckett, R. J., Knight, T. C. D., and Cook, W. S.
    (1976). *Science 196*, 786.

Owen, T. (1974a). *Comments Astrophys, Space Phys.* 5, 175.

Owen, T. (1974b). *Science 183*, 763.

Owen, T. (1976). *Icarus 28*, 171.

Hugenin, R. L. (1976). *Icarus 28*, 203.

Masursky, H. (1973). *J. Geophys. Res.* 78, 4009.

Milton, D. J. (1973). *J. Geophys. Res. 78*, 4037.

Sagan, C., Toon, O. B., and Gierasch, P. J. (1973). *Science
    181*, 1045.

11

SEARCH FOR ORGANIC AND VOLATILE INORGANIC COMPOUNDS IN TWO SURFACE SAMPLES FROM THE CHRYSE PLANITIA REGION OF MARS

K. Biemann, J. E. Biller
Department of Chemistry, Massachusetts Institute of Technology
Cambridge, Massachusetts

J. Oro
Department of Biophysical Sciences
University of Houston, Houston, Texas

L. E. Orgel
Salk Institute for Biological Studies
San Diego, California

A. O. Nier
School of Physics and Astronomy
University of Minnesota, Minneapolis, Minnesota

D. M. Anderson
Office of Polar Programs
National Science Foundation, Washington, D.C.

P. G. Simmonds
Organic Geochemistry Unit, School of Chemistry
University of Bristol, Bristol, England

D. Flory

Spectrix Corporation, Houston, Texas

A. V. Diaz

NASA Langley Research Center,
Hampton, Virginia

D. R. Rushneck

Interface, Inc., Fort Collins, Colorado

*Two surface samples collected from the Chryse Planitia
region of Mars were heated to temperatures up to 500°C, and
the volatiles that they evolved were analyzed with a gas chromatograph-mass spectrometer. Only water and carbon dioxide
were detected. This implies that organic compounds have not
accumulated to the extent that individual components could be
detected at levels of a few parts in 10⁹ by weight in our samples. Proposed mechanisms for the accumulation and destruction
of organic compounds are discussed in the light of this limit.*

The objective of the Viking molecular analysis experiment
is to analyze periodically the composition of the atmosphere at
the surface of Mars and to search for organic compounds and
certain inorganic volatiles in the surface material at the
landing site. The principal instrument used for both investigations is a mass spectrometer (MS). It is used in conjunction
with a gas chromatograph (GC) for the organic analyses. The
objective of the entire system has been described previously
(Anderson et al., 1972; Biemann, 1974) as have the results of
the atmospheric analyses (Owen and Biemann, 1976; Biemann et
al., 1976).

A gas chromatograph-mass spectrometer (GCMS) was chosen for
the search for organic compounds chiefly because of its sensitivity and versatility. A number of mechanisms could contribute
to the contemporary accumulation of organic compounds on Mars,
for example, photochemical or biological processes and meteoritic infall. It is also possible that fossil organic compounds

are present in the soil which reflect synthesis under conditions which existed on Mars long ago. The MS could analyze mixtures formed in any of these ways, and it was hoped that the details of the analyses would help to elucidate the mechanisms of synthesis.

Because the experiment involves the expulsion of the volatile or pyrolyzable constituents of the surface sample, one obtains, incidentally, information about volatiles such as water or CO_2 that might be released from the mineral matrix.

EXPERIMENTAL DETAILS

The instrument and some of the performance data have been described previously (Anderson et al., 1972; Biemann, 1974). Briefly, the instrument consists of a set of three small ovens in which a crushed surface sample (particle size <300 μm) can be heated for 30 seconds to any of three temperatures ($200°$, $350°$, or $500°C$) to expel volatile substances and pyrolysis product. These substances are then swept onto the GC column[1] by a stream of $^{13}CO_2$[2], and eluted from the column with H_2 being used as a carrier gas.

The GC column temperature is held at $50°C$ for the first 10 minutes, then linearly increased to $200°C$ in 18 minutes and held at this temperature for another 18, 36, or 54 minutes as desired. (Any of these three time periods can be selected by

[1] The GC column is filled with a liquid-modified organic adsorbent consisting of 60- to 80-mesh Tenax-GC (2,6-diphenyl-p-phenylene oxide coated with polymetaphenoxylene. This specific packing was developed to (i) maximize the separation of H_2O and CO_2 from organic compounds, (ii) transmit efficiently most compound classes at the low nanogram level, (iii) have exceptional thermal stability, and (iv) have mechanical strength compatible with the rigors of space flight (Novotny et al., 1975).

[2] The use of ^{13}CO for flushing the oven during the heating period was a late change in design. It was necessitated by the observation that some compounds were reduced by the hydrogen previously used in the step. Isotopically labeled CO_2 was chosen in order to distinguish it from any CO_2 involved from the sample upon heating. Furthermore, it would make it possible to detect incorporation of CO_2 into organic compounds during the heating or pyrolysis.

ground command.) The carrier gas is removed by passage through
a palladium separator (Lucero and Haley, 1968; Simmonds *et al.*,
1970; Dencker *et al.*, 1972), and the eluting components enter
the MS.

To protect the MS from an excess of material that could
harm the ion source or pump, the latter automatically controls
an effluent divider which reduces in steps the fraction of the
effluent that enters the separator, and vents the remainder of
the effluent to the atmosphere (Anderson *et al.*, 1972; Biemann,
1974).[3]

The use of the effluent divider to protect the instrument
results in a decrease of sensitivity in each segment of the
chromatogram, this sensitivity being inversely proportional to
the split ratio of the effluent divider operative in that segment.

The MS repetitively and continuously scans from *m/e* 12 to
200 in 10 seconds and has a dynamic range of $1:10^7$. Each spectrum is recorded as 3840 digital data points, each one representing an ion current value encoded to 9 bits on a log scale.
All these data, together with various other operating parameters, are telemetered back to Earth (about 1.4 to 1.8 x 10^7 bits
per experiment). The data are processed on the ground by means
of computer techniques that were developed previously (Hertz *et
al.*, 1971) and were adapted specifically for the particular
application discussed (Biller and Biemann, 1977).

After the Viking I spacecraft was launched on 20 August
1975, the instrument was tested during the Earth to Mars cruise
from November 1975 to January 1976. One of the more important
tests consisted of a complete "blank" experiment during which
one of the sample ovens was heated to 500°C. The data (504
mass spectra and all associated engineering data) were returned

[3]*This "effluent divider" can be preset in two different
modes so as to provide maximum protection for the MS. In the
hydrous mode, the assumption is that any overload is due to a
very wide GC peak, for example, from water. In this mode, in
order to avoid many rapid changes of state that would confuse
data interpretation a logical time filter count is utilized,
which inhibits changes of state to a lower state of division for
longer periods of time than in the anhydrous mode. The following indicates the relationship of the four vent-ratios and the
time filter count for changes to the next lower state in the
hydrous mode with the count for the anhydrous mode in parentheses: 8000/1, 45 sec (45 sec); 400/1, 2 minutes (45 sec);
20/1, 15 minutes (45 sec); 3/1, 15 minutes (45 sec).*

to Earth and analyzed. They confirmed the presence of a few homologous oligomers of fluoropropyleneoxide (Freon-E type) in addition to residual adsorbed water. The level of contamination was sufficiently low that it did not cause concern. In fact, the impurities, fortuitously, provide an excellent mass standard.

After the successful landing of Viking 1, three separate samples of the martian surface were acquired by the surface sampler on sol 8 (GCMS-1), sol 14 (GCMS-2), and sol 31 (GCMS-3). The sol 8 acquisition was predominantly a fine-grained subsurface sample (from a depth of 4 to 6 cm) mixed with, at most, 10 percent of surface material. This sample was not immediately analyzed as planned, because the processor and distribution assembly (PDA) failed to indicate a full sample cavity. However, in view of the small amount of material required (<200 mg) to fill one of the ovens, the decision was made to command the GCMS to proceed with the first analysis of the sol 8 sample at 200°C, while attempting to acquire a second sample.

The next sample, acquired on sol 14, was also a subsurface sample and was collected directly adjacent to the sol 8 site. The sol 14 (GCMS-2) sample was never analyzed but was discarded in favor of a sample from a very different site. The sol 31 (GCMS-3) sample was acquired in an area about 3 m from the sol 8 sites and was primarily a surface sample, composed of a course cohesionless granular material. A more complete description of the sites and the physical properties of the surface material are given elsewhere (Shorthill *et al.*, 1976). The sample sites, times of acquisition, and conditions of analysis for the two samples from the Chryse Planitia region are summarized in Table I.

RESULTS

During the analysis of the first sample, traces of methylchloride and perfluoroethers of the Freon-E type were detected. These are contaminants previously encountered in preflight and cruise tests. Their detection as sharp gas chromatographic peaks producing the proper mass spectra demonstrates the correct functioning of all parts of the instrument. None of the five gas chromatograms obtained in the experiments listed in Table I showed any indication of the presence of organic compounds indigenous to the two samples. However, at 350 and 500°C considerable quantities of water were evolved (Table II).

In experiments with the second sample we did not detect methylchloride or perfluoroether contaminants. Water eluting from the column, and accumulated water background in the instrument kept the effluent divider in a higher ratio than previously, thus obscuring the small amounts of these compounds that were present.

TABLE I. *Acquisition Sites and Conditions of Analysis for the Two Maritan Samples*

Date of Analysis	Temperature (^{O}C)	Mode	Time held at temperature
Sample 1(GCMS-1 (9)) subsurface; acquired on sol 8 (29 July 1976)			
Sol 17 (7 August)	200°	Hydrous	18
Sol 23 (13 August)	500°	Anhydrous	36
Sample 3(GCMS-2 (9))surface; acquired on sol 31 (21 August 1976)			
Sol 32 (22 August)	350°	Hydrous	54
Sol 37 (27 August)	500°	Hydrous	54
Sol 43 (2 September)	500°	Hydrous	36

The absence of evidence for organic compounds in the data derived from these two samples from the surface at Chryse Planitia should be discussed. Obviously, compounds that are not transmitted by the GC column[1] or its interface with the MS cannot be detected. The detection limit for compounds that are transmitted can be calculated from the mass spectral data obtained in these experiments. Since the retention behavior of any compound can be predicted, we examined the signal observed at the proper retention time in the mass chromatogram (Hertz *et al.* 1971) and used the intensity of the most diagnostic, intense ion signal of the compound in question to estimate its abundance. In no case of interest was there a maximum in the mass chromatogram; this indicates the absence of all the compounds which we have searched for as yet, at the level corresponding to the background ion current. Table III lists the upper limits for a few typical compounds. A range is given because the sensitivity differs from run to run due to the variability of the effluent divider state. It should be noted that our estimates represent upper limits at the present state of data reduction, and do not imply that we believe that the substances referred to in Table III are present at that or a lower level. They were selected solely as examples because they have been detected at a level of 0.01 to 10 parts per million (ppm) in other relevant materials, for example, a terrestrial soil from the Antarctic (Quam, 1971; Cameron *et al.*, 1970) and the Murchison meteorite, by means of an instrument virtually identical with that operating on the surface of Mars (A. L. Lafluer, unpublished results).

There are a number of small molecules of intrinsic interest in relation to nonbiological as well as biological chemis-

TABLE II. Results of the Molecular Analysis Experiments

Material	Quantity (temperatures in degrees Celsius)
I. Inorganic	
Carbon dioxide	Some in all experiments (quanitation not yet available)
Water	Sample 1: at 200^0, much less than 0.1%
	at 500^0, 0.1 to 1.0%
	Sample 2: at 350^0, 0.1 to 1.0%
	at 500^0, somewhat less than at 350^0
II. Organic	None detected (see Table III for detection limits)
III. Terrestrial contaminants	
Methyl chloride	~ 15ppb
Fluoroethers	1 to 50 ppb

TABLE III. *Upper limits of selected organic compounds which would be detected if present, in the two samples from the Chryse Planitia region of Mars*

Compound	Range of detection limits (parts per 10^9)
Alphatic hydrocarbons	
Butene	< 1 to 10
Hexane	< 1 to 10
Octane	< 1 to 10
Aromatic hydrocarbons	
Benzene	0.5 to 5
Toluene	0.5 to 5
Naphthalene	0.05 to 0.5
Oxygen-containing compounds	
Acetone	< 10 to 50
Furan	< 0.1 to 1
Methylfuran	< 0.2 to 2
Nitrogen-containing compounds	
Acetonitrile	< 1 to 10
Benzonitrile	< 0.2 to 2
Sulfur-containing compounds	
Thiophene	< 0.1 to 0.5
Methylthiophene	< 0.1 to 0.5

try for which it is difficult to set a detection limit because of (1) the way in which the instrument was operated, (2) the relatively large amount of water that was evolved at temperatures higher than 200°C, and (3) the use of $^{13}CO_2$ to flush the vaporized material out of the sample oven and onto the GC column. For this reason, compounds like CH_4, C_2H_2, C_2H_4, C_2H_6, and probably NH_3 would have been completely obscured. For other very volatile materials such as CH_3OH, CH_2O, HCN, $(CN)_2$, and C_3H_{4-8} the detectabiliy is limited by the split ratio of the effluent divider in the relavent part of the chromatogram, particularly where CO_2 and H_2O elute. The limit of detection estimated for these compounds is in the tens of parts per million for most experiments. One exception is the 200°C experiment with sample 1, where the split ratio in this region of the gas chromatogram was much more favorable (3:1, that is, 25 percent of the effluent admitted to the MS), in which case this group would have been detected at levels of tens of parts per billion.

In view of the results obtained by Viking 1 it is planned to modify the strategy of the experiments to be conducted with the instrument on Viking lander 2 in such a manner as to eliminate the contributions of the $^{13}CO_2$ to the obscuration of the most volatile components that might evolve from the sample upon heating.

We have not been able to detect SO_2, free sulfur, or H_2S. While traces of H_2S would have been vented with the CO_2, thus escaping detection, any appreciable quantities of either S or H_2S entering the palladium separator would have substantially decreased its efficiency or even inactivated it completely. The efficient operation of the separator provides indirect evidence that these two substances were not evolved at levels corresponding to a few parts per million in the samples.

DISCUSSION

The failure to find organic compounds at the detection limit of our instrument in the two Viking lander 1 sampling sites is an important observation that must be carefully evaluated. It has implications concerning the composition of the planet in the Chryse Planitia region, the fate of organic compounds from nonmartian sources falling on the planet's surface, and the steady-state concentration of any organic compounds that are being synthesized by photochemical processes. It also provides an upper limit to the amount of biologically maintained organic material in the samples.

Cosmochemical considerations and studies on the volatiles of Mars (Levine, 1976; Owen, 1976; Oro, 1976) indicate that substantial quantities of CO_2 must have been derived from more reduced forms of carbon during the planet's history. The fact that such carbon compounds are not present in detectable levels in the Chryse Planitia region does not necessarily preclude their existence in less disturbed areas of the planet's surface or interior, where they may have been protected from oxidation, photo-destruction, and other degradative processes.

The contribution of organic compounds to the martian regolith from nonmartian sources (mainly meteoritic matter can be calculated from estimates of their influx rate on Mars relative to that on the moon. It is estimated that after mixing in a regolith that has a mean thickness of 4.6 m, meteoritic input at three Apollo sites on the moon contributes an amount of material equivalent to 1.1% of type 1 carbonaceous chondrites (or C-1 equivalent component) to the regolith (Morgan, 1976). The infall of meteorites on Mars has been estimated by some authors to be approximately twice that on the moon (Soderblom *et al.*, 1974). The martian regolith is considered to be approximately 2 km deep (Fanale, 1976), that is, several orders of magnitude larger than the regolith on the moon. Thorough mixing with the regolith would cause substantial dilution of the meteoritic input of organic compounds, and their concentrations would be reduced to an equivalent of about 0.005% of material from type 1 carbonaceous chondrites. Therefore, an organic compound, such as naphthalene, detected in the Murchison meteorite at the level of 1 ppm with the laboratory version of the Viking lander 1 instrument (A. L. LaFleur, unpublished results) would, according to these estimates be diluted in the martian regolith to 0.05 part per billion (part per 10^9).

However, some authors suggest a higher meteoritic infall rate on Mars (Anders and Arnold, 1965), and a mixed regolith layer of only about 100 m. If this were correct the abundance of naphthalene in the martian soil would be about 10 ppb, which would be detectable by the instrument on Mars. The fact that we have not found naphthalene at this level argues against such a model unless the meteoritic organic compounds have been destroyed to a large extent.

Contemporary synthesis of organic compounds on the surface of Mars, or on dust particles suspended in its atmosphere, represents another source of such substances. Hubbard *et al.* (1973) have demonstrated the synthesis of organic compounds by an ultraviolet light-induced reaction of CO with water adsorbed on inorganic matrices under simulated martian conditions. The steady-state concentrations of the primary products of this process are just below our detection limit[4]. The fact that we have

---

[4] *Calculations on the yield of photochemically induced organic compounds give an upper limit of 100 nmol/gm of martian soil (J.S. Hubbard, personal communication). It is further estimated that about 20% is recoverable as an organic fraction with the approximate distribution: $HCOO^3M$, 20 nmol; CH_2O, 0.5 nmol; CH_3CH), 0.25 nmol; $HOCH_2COOH$, 0.5 nmol. For the 100-mg sample used in our experiments this would correspond to concentrations of 920, 15, 11, and 38 ppb, respectively. Formate and glycolic acid would probably decompose during analysis; the high detection limit for formaldehyde has been discussed in the text. Acetaldehyde at a maximum concentration of 11 ppb is just below the minimum detection limit, provided most or all was admitted to the mass spectrometer, that is, at a low affluent MS, that is, at a low effluent divider state.*

not found organic compounds implies that the accumulation of stable products by further condensation reactions of the primary photochemical products has not been a major process.

Finally, the implications of our results on the interpretation of direct life-detection experiments should be discussed. The Viking GCMS was not intended to be an instrument for the detection of life. However, if organic compounds were found in the martian soil, such a finding could be considered under certain circumstances supportive of the positive identification of biological processes on Mars. The reverse is not necessarily true, however. It is possible to have a small population of microorganisms in a sample without being able to detect the organic compounds evolved from their biomass[5].

Substantial amounts of residual organic matter are usually found associated even with small numbers of microorganisms on Earth. More than 20 organic compounds were identified, at the level of 0.01 to 1 ppm, in a 100-mg sample of Antarctic soil (A. L. LaFleur, unpublished results) that was supposed to contain less than 10^4 microorganisms per gram (Quam, 1971; Cameron et al., 1970). If substantial numbers of organisms were present in our sample they would have to be much more efficient than terrestrial organisms as scavengers of biological debris.

Having discussed the possible sources of organic materials we should also mention potential degradative processes that could cause their disappearance. It is difficult to estimate the degree to which oxidation and other degradative processes might reduce the abundance of organic compounds in the martian regolith. Possible destructive agents include ultraviolet radiation, oxygen, and such oxidants as nitrates and metal oxides, acting independently or synergistically. Certainly, continuous exposure to short-wave ultraviolet light, in the presence of oxygen, will cause rapid photolysis of most organic compounds. Even ultraviolet light of longer wavelength is known to photooxidize many organic compounds on the surface of certain metal oxide catalysts (for example, FeO, Fe_2O_3, and TiO_2) (Formenti et al., 1971).

---

[5] The average detection limit (Table II) of the GCMS for a single compound is 1 ppb (10^{-2} gm/gm of soil). If we assume that the efficiency of pyrolysis is 10%, then 10^{-7} gm of organic matter would be required for the detection of a major component that represented 10% of the total pyrolyzate-volatile fraction. Furthermore, if the dry weight of a typical prokaryotic cell were taken as 10^{-13} gm, then 10^6 organisms would be required to yield 10^{-7} gm of organic matter.

The destructive oxidation of organic compounds by nitrate, either in situ or during the heating of the martian sample in the GCMS oven, should also be considered. However, limited experiments with terrestrial soils (Horowitz *et al.*, 1969) indicate that only a fraction of the organic matter is oxidized when samples are heated to 500°C in the presence of excess nitrate.

The presence of highly oxidizing inorganic constituents in the martian surface material has been inferred from some of the data generated by two of the Viking biology experiments (Klein *et al.*, 1976). For a number of reasons, we tend to rule out the possibility that such agents, if present, have destroyed organic material in our sample during the 30-second heating period. First, the oxygen evolution, 725 nmole of O_2 per cubic centimeter of surface sample, would at most oxidize the equivalent of 6 μg of reduced carbon, if the oxidation were quantitative; second, the evolution of water, even at 200°C, is sufficient to convert the oxidizing material to O_2; and third, the observation of CH_3Cl in one of the experiments seems to demonstrate that organic compounds are not oxidized to the extent that they would escape detection.

MINERALOGICAL IMPLICATIONS

The data of particular interest from a mineralogical point of view are those dealing with the evolution of volatiles produced by thermal destruction of minerals such as hydrates or hydroxides, carbonates, sulfates, and nitrates. Data of the lander imagery (Mutch *et al.*) and inorganic chemical (Toulmin *et al.*, 1976) investigations strongly suggest the presence of fine-grained ferric oxide or hydroxide in the material sampled, but do not definitely discriminate between anhydrous ferric oxide (presumably hematite) and hydrated forms, the commonest of which, at least terrestrially, is goethite, FeOOH. The data presented here indicate that most of the water is evolved at 350°C heating. This behavior seems likely to reflect the presence of some hydrate other than goethite. Direct experimental data on decomposition of minerals at the heating rates and durations used in the present experiments are scanty, and the dehydration temperature of goethite is known to depend on many factors, such as degree of crystallinity, which cannot be (Pollack *et al.*, 1970a,b) determined in this instance. Calculations from kinetic data indicate, however, that the characteristic (e^{-1}) time for dehydration of goethite should be tens of days at 200°C, tens of minutes at 350°C, and a few seconds at 500°C. If these figures are of the correct order, the H_2O evolved from the martian samples could not have been derived from goethite. Hydrated sulfates may be present in the martian surface material (24), and could possibly be the sources of the H_2O evolved between 200° and 350°C.

Most terrestrial clay minerals lose their lattice H_2O at temperatures above 500°C; physically adsorbed or loosely bound H_2O, if present, would not have been detected as it would have been desorbed between sample acquisition and analysis (Anderson *et al.*, 1967). Thus clay minerals should not have contributed to the water that we detected. Detailed testing of various materials on the laboratory GCMS may permit more confident identification of the hydrates in our samples.

The very low upper limit on evolved sulfur corresponds to partial pressures of the order of 10^{-4} atm (maximum) in the oven at 500°C. This value does not preclude a priori the existense of mineral sulfides in the sample. On the other hand, terrestrial soils containing sulfide minerals have evolved detectable amounts of reduced sulfur species in the laboratory instrument, and we, therefore, regard the presence of sulfides (other than extremely refractory ones, for example, troilite) as very unlikely.

The data also suggest that the sample evolved CO_2 during heating, but complexities in the dynamics of flow within the instrument make interpretation difficult at this time. More detailed analysis of the data, and testing on the laboratory instrument, may enable us to make better estimates of the amounts of CO_2 evolved, and hence better understand the nature of the carbonates present in the samples.

CONCLUSIONS

Our preliminary results obtained from two samples of the surface of Mars in the Chryse Planitia region indicate that the material releases only 0.1 to 1% of water when it is heated to 350° or 500°C. This water is presumably present in mineral hydrates that release water at these high temperatures. The samples do not contain organic compounds more complex than, for example, propane or methanol, at the level of a part per 10^9. Such low-molecular-weight compounds were not found at detection limits in the range of 1 part per 10^5. These results seem to exclude the existence of any efficient contemporary process that produced organic compounds. It also makes it unlikely that there are low-efficiency processes occurring over a long period, which slowly accumulate organic compounds stable in the martian environment. It has to be reiterated that these findings relate to only two samples at one landing site, and care has to be taken in extrapolating our conclusions to other parts of the planet. It is hoped that the results soon to be obtained from the Viking lander 2 will add another dimension to our knowledge.

ACKNOWLEDGMENTS

We are indebted to all of the many individuals whose efforts and talents have contributed to this work. We thank specifically H. C. Urey, G. A. Shulman and R. A. Hites for their contributions at the earlier phases of the Viking activities; A. La Fleur, J. Lavor, and E. Ruiz who conducted and processed most of the tests on the laboratory instrument, our fellow team member F. Owen for his continuous interest and stimulation; D. Howarth, J. Rampacek, R. Williams, and C. Hang for their unfailing expert assistance before and during the mission, and finally L. Crafton without whose labor and prodding this manuscript could not have been produced. This work was supported by NASA Research Contract NAS 1-9684 and many other related ones. P.G.S. is a research fellow at the University of Bristol.

REFERENCES

Anders, F., and Arnold, J. R. (1965). *Science 149*, 1494.
Anderson, D. M., Goffney, E., and Low, P. F. (1967). *Science 153*, 319.
Anderson, D. M., Biemann, K., Orgel, L. E., Oro, J., Owen, T., Shulman, G. P., Toulmin, P. III, and Urey, H. C. (1972). *Icarus 16*, 11.
Biemann, K. (1974). *Origins Life 5*, 417.
Biemann, K., Owen, T., Rushneck, D. R., LaFleur, A. L., and Howarth, D. W. (1976). *Science 194*, 76.
Biller, J. A., and Biemann, K. In preparation.
Cameron, R. E., King, J., and David, C. N. (1970). *Soil Sci. 109*, 110.
Dencker, W. D., Rushneck, D. R., and Showmake, G. R. (1972). *Anal. Chem. 44*, 1753.
Fanale, F. P. (1976). *Icarus 28*, 179.
Formenti, M., Juillet, F., Meriandeau, P., and Teichner, S. J. (1971). *Chem. Technol. 1*, 680.
Hertz, H. S., Hites, R. A., and Biemann, K. (1971). *Anal. Chem. 43*, 681.
Horowitz, N. H., Bauman, A. J., Cameron, R. E., Geiger, P. J., Hubbard, J. S., Shulman, G. P., Simmonds, P. G., and Westberg, K. (1969). *Science 164*, 1054.

Hubbard, J. S., Hardy, J. P., Voecks, G. E., and Golub, E. E. (1973). *J. Mol. Evol. 2*, 149.

Klein, H. P., Horowitz, N. H., Levin, G. P., Oyama, V. I., Lederberg, J., Rich, A., Hubbard, J. S., Hobby, G. L., Straat, P. A., Berdahl, B. J., Carle, G. C., Brown, F. S., and Johnson, R. D. (1976). *Science 194*, 99.

Levine, J. S. (1976). *Icarus 28*, 165.

Lucero, D. P., and Haley, F. C. (1968). *Gas Chromatogr. 6*, 477.

Morgan, J. W. (1976). *Proc. Internat. Conf. Modern Trends in Activation Analysis 1976, Munich, West Germany.*

Mutch, T. A., Arvidson, R. A., Binder, A. B., Huck, F. O., Levinthal, E. C., Liebes, S. Jr., Morris, E. C., Nummedal, D., Pollack, J. B., and Sagan, C. (1976). *Science 194*, 87.

Novotny, M., Hayes, J. M., Bruner, F., and Simmonds, P. G. (1975). *Science 189*, 215.

Oro, J. (1972). *Space Life Sci. 13*, 507.

Owen, T. (1976) *Icarus 28*, 171.

Owen, T., and Biemann, K. (1976). *Science 193*, 801.

Pollack, J. B., Pitman, D., Khare, B. N., and Sagan, C. (1970a). *J. Geophys. Res. 75*, 2480.

Pollack, J. B., Wilson, R. N., and Goles, G. N. (1970b). *J. Geophys. Res. 75*, 7491.

Quam, L. O. (ed) (1971). "Research in the Antarctic," pp. 137-189. AAAS, Washington, D. C.

Shorthill, R. W., Hutton, R. E., Moore, H. J., Liebes, S., Scott, R. F., and Spitzer, C. R. (1976). *Science 194*, 91.

Simmonds, P. G., Shoemake, G. R., and Lovelock, J. E. (1970). *Anal. Chem. 42*, 881.

Soderblom, L. A., Condit, C. D., West, R. A., Herman, B. M., and Kreidler, T. J. (1974). *Icarus 22*, 239.

Toulmin, P. III, Clark, B. C., Baird, A. K., Rose, H. J. Jr., and Keil, K. (1976). *Science 194*, 81.

12

THE ATMOSPHERE OF MARS
NEAR THE SURFACE:
ISOTOPE RATIOS AND UPPER LIMITS
ON NOBLE GASES

K. Biemann, A. L. LaFleur

Department of Chemistry, Massachusetts Institute of Technology
Cambridge, Massachusetts

T. Owen

Department of Earth and Space Sciences
State University of New York, Stony Brook, New York

D. R. Rushneck

Interface, Inc., Fort Collins, Colorado

D. W. Howarth

Guidance and Control Systems Division
Litton Industries, Woodland Hills, California

Several new analyses of the martian atmosphere have been carried out with the mass spectrometer in the molecular analysis experiment. The ratios of abundant isotopes of carbon and oxygen are with 10% of terrestrial values, whereas nitrogen-15 is considerably enriched on Mars. We have detected argon-38 and set new limits on abundances of krypton and xenon. The limit on krypton is sufficiently low to suggest that the inventories of volatile substances on Mars and on Earth may be distinctly different.

213

We have obtained new data on the composition of the martian atmosphere using the mass spectrometer of the molecular analysis experiment (Owen and Biemann, 1976). The purpose of this set of investigations was to determine the relative abundances of the isotopes of argon, carbon, oxygen, and nitrogen and to search for trace constituents, especially the other noble gases. Analysis of these results and the acquisition of additional data are continuing.

A summary of the analyses carried out is given in Table I. Atmospheric gases were accumulated in the sample chamber (1) as unaltered atmosphere for direct analysis, or (2) as enriched atmosphere in which progressively admitted samples of the atmosphere were subjected to treatment with Ag_2O, and LiOH (as described Owen and Biemann (1976) to absorb the CO and CO_2, and $Mg(ClO_4)_2$ to remove the resulting water, so that the partial pressures of the trace gases would be increased before the contents of the sample chamber were analyzed. In this way, the cycling and chemical scrubbing of ten samples of martian atmosphere admitted progressively to the analytical chamber led to a sevenfold enrichment of minor constituents; the addition of nine more sampling and adsorption cycles produced a final enrichment factor of 8.5.

Two excitation voltages were used to help discriminate among doubly and singly charged ions, and the enriched samples were always analyzed at both voltages. The first voltage given in Table I is that at which an instrumental background was also recorded.

The results derived from this series of experiments to date are summarized in Tables II and III. The isotope ratios for carbon and oxygen in Table II are based on the relative inten-

TABLE I. Atmospheric Analyses on Mars

Date (August 1976)	*Sample*	*Number of analyses*	*Excitation voltage (electron volts)*
6 to 7	*Unaltered*	*7*	*70*
14	*Enriched (10 cycles)*	*2*	*70 and 45*
16	*Same as above, reanalyzed*	*2*	*45 and 70*
23	*Enriched (19 cycles)*	*2*	*45 and 70*

sities of $^{12}C^{16}O^{16}O$, $^{13}C^{16}O^{16}O$, and $^{12}C^{16}O^{18}O$ in the analyses
of unaltered atmosphere; they show no significant departure
from average terrestrial values. There is a definite enhancement of the abundance of ^{15}N relative to ^{14}N on Mars, but we
are unable to be precise in our evaluation of the ratio of
these isotopes because of possible interference at m/e 29 from
^{13}CO desorbed or formed in the instrument. (We cannot use m/e
15 as a check because of background interference.) $^{13}CO_2$ is
introduced into the mass spectrometer in the organic analyses
(Biemann et al., 1976) and we can see its effects at m/e 29 in
the analyses of unaltered atmosphere. We must therefore make
a correction for the worst case of contamination on the assumption that this same contamination occurs in the mass spectra of
the enriched samples, because all of the enriched samples were
analyzed after an organic analysis had been performed. The
range given in Table II reflects the value derived without
correction (^{15}N ^{14}N = 0.0064, which is a 74 percent enhancement
of ^{15}N) and the value obtained after correcting for the worst
case of contamination described above ($^{15}N/^{14}N$ = 0.0050, 36
percent enhancement of ^{15}N). Our results therefore just em-
brace the value derived by Nier et al. (1976) who have discussed the significance of this finding.

We have detected ^{38}Ar in the martian atmosphere and we can
evaluate its abundance relative to ^{36}Ar, which we reported previously (Owen and Biemann, 1976). Unfortunately, the instrumental background in this region has deteriorated and this,
with the memory effects from the ion pump (release of previously buried argon), has prevented an accurate specification of
the relative abundances of these two isotopes. The range given
in Table II encompasses the terrestrial atmospheric value.

We have searched unsuccessfully for methane, neon, krypton,
and xenon (Table III). The methane limit is extremely high because of the substantial background in the instrument at the
low masses. A global limit of 25 parts per billion of methane
in the martian atmosphere has been established by infrared
spectroscopy from Mariner 9 (Maguire, 1977). Our upper limit
for neon is set by the presence of doubly ionized ^{40}Ar at m/e
20 in the low voltage analyses. Thus, in contrast to the other
noble gases, the enrichment process does not lower our previous
limit. The other common isotope of neon at m/e 22 is expected
to be on the order of one-tenth as abundant as ^{20}Ne and our maximum enrichment has not brought it above the threshold set by
CO_2^{2+}.

The enrichment process considerably lowered the limits for
krypton and xenon compared to those derived previously (Owen
and Biemann, 1976). The expected values for the krypton and
xenon isotopes given in Table III are based on terrestrial
abundances scaled from ^{36}Ar. It is evident that we have

TABLE II. *Comparison of the Isotope Ratios in the Martian and Terrestrial Atmospheres*

Species	Mars	Earth
$^{15}N/^{14}N$	0.0064 to 0.0050	0.00368
$^{13}C/^{12}C$	0.0118 $\pm$ 0.0012	0.0112
$^{18}O/^{16}O$	0.00189 $\pm$ 0.0002	0.00204
$^{86}Ar/^{38}Ar$	4 to 7	5.3

TABLE III. *Upper Limits (Expressed as Parts Per Million (ppm)) of Gases Not Detected in the Martian Atmosphere*

Species	Present upper limit (ppm)*	Expected value (ppm)
CH_4	<120	<25†
Ne	<10	5
Kr	<0.3	0.3
Xe	<1.5	0.02

*Based on calibration with pure gases of terrestrial isotope distribution.

†Value for CH_4 expressed as parts per billion.

reached the level of sensitivity at which we would expect to detect krypton on Mars if it were present in the terrestrial abundance ratio.

These results place us in an interesting dilemma. The relatively low abundance of ^{36}Ar on Mars suggests that the total outgassing of the planet was perhaps as little as 1/100 the amount exhibited by Earth. But the enhancement of ^{15}N relative to ^{14}N implies that a large amount of nitrogen has escaped, which would indicate a much larger degree of total outgassing (Nier and McElroy, 1976). However, these inferences are made by assuming that the original volatile inventory on Mars was assentially identical to that on Earth.

If, on the contrary, the martian volatile inventory was more primitive, that is, closer to that exhibited by the type I carbonaceous chondrites, one could accommodate an initial nitrogen abundance up to 30 times the present atmospheric value, while increasing the carbondioxide abundance only by a factor of 10. This would be consistent with the previous low outgassing model in which the maximum surface pressure was on the order of 100 mbar (Owen and Biemann, 1976).

Another possibility is that sweeping by the solar wind has drastically reduced the ^{36}Ar from an earlier amount that corresponded to a massive primitive atmosphere with terrestrial relative abundances. But this hypothesis would imply some enrichment of ^{84}Kr relative to ^{36}Ar in the present atmosphere, which we do not observe. On the other hand, an inventory of the C-1 type postulated above would predict a lower $^{84}Kr/^{36}Ar$ ratio on Mars than we find in the atmosphere of Earth. This seems to be the case.

It is too early to discriminate among these various possibilities. We anticipate that the mass spectrometer on Lander 2 will have better sensitivity for atmospheric analyses, based on calibrations before launching and on backgrounds obtained during cruise. We can also improve the enrichment factor at some risk to the instrument after completing all of the other types of analyses. We thus hope to be able to improve our existing precision on the isotope ratios and to lower still further the threshold for detecting trace constituents.

ACKNOWLEDGMENTS

We thank our fellow team members, D. Anderson, L. Orgel, J. Oro, P. Toulmin III, and especially A. O. Nier, for helpful discussions and enthusiastic support. We are indebted to J. A. Biller, L. Crafton, A. V. Diaz, E. M. Ruiz, and R. Williams for essential contributions to the success of the experiment. This work was supported by NASA research contracts NAS 1-10493 and NAS 1-9684.

REFERENCES

Biemann, K., Oro, J., Toulmin, P. III, Orgel, L. E., Nier, A.
    O., Anderson, D. M., Simmonds, P. G., Flory, D., Diaz, A.
    V., Rushneck, D. R., and Biller, J. A. (1976). *Science 194*,
    72.
Maguire, W. C. (1977). *Icarus,* in press.
Nier, A. O., McElroy, M. B., and Ynug, Y. L. (1976). *Science
    194*, 68.
Owen, T., and Biemann, K. (1976). *Science 193*, 801.

13

WATER IN THE MARTIAN REGOLITH

Duwayne M. Anderson

Division of Polar Programs, National Science Foundation
Washington, D.C.

Water is a known constituent of the Martian atmosphere. It
is present near the surface at partial pressures near saturation.
This is prima facie evidence of its presence in the Martian
regolith. Possible forms include free water in the form of
ice; very concentrated brines or water of deliquesence; adsorbed, interlayer, and zeolitic water; crystalline hydrates;
and silicate lattice water. The first three categories are of
special significance in considering the possibility of the
existence of microorganisms in the regolith. The amounts and
forms of water in the Martian regolith are unknown at present
and are subjects for speculative discussion. This chapter
considers the evidence bearing on this point, including the
qualitative result of the first Viking I GCMS analysis of the
aeolian material at the Chryse landing site.

For some time there has been no doubt that water exists
on Mars. It has been confirmed as a constituent of the Martian
atmosphere in quantities approaching saturation. The radiometric temperatures measured by the Mars Mariner Missions established the fact that Mars is a permafrost planet. Ice and
water vapor are the stable phases of water. The liquid, under
the present climatic regime can be only an ethemeral phase.
Other forms of water likely to be found include adsorbed phases,
interlayer, and zeolitic water found in clay minerals, crystalline hydrates, and other combined forms. Free water can
exist in the form of ice, and very concentrated brines and
waters of deliquescences are also possible. The role of interfacial water in the early stages of biochemical evolution was
discussed earlier by Anderson and Banin (1975).

The marked hiatus in the observed retreat of the Martian polar cap combined with the radiometric temperatures observed by Mariner 9 is consistent with the conclusion that the Martian polar caps are mainly composed of water ice (Murray and Malin, 1973; Soderblom *et al.*, 1973). Frost and low lying clouds having temperatures indicative of water ice have been observed by both the Mariner and Viking orbitors (Briggs, 1976). This has led to the view that water ice is a common constituent of the Martian regolith where it may have created an ice-cemented permafrost.

Terrestrial permafrost is defined with reference to ground temperature only. Ice content is considered only in the classification of permafrost into such catagories as unfrozen, dry, or ice rich. In common usage the term is often taken to indicate the presence of ground ice. Because inevitably it is necessary to make distinctions, when one considers the situation on Mars there seems to be no reason to depart from conventional terminology. Thus, the temperature regime of the Martian regolith is the principle concern.

Pursuing this point further, the essential aspects of the permafrost conditions are illustrated in Fig. 1 and 2. Figure 1 shows the near surface thermal regime. Points a and b are the mean annual minimum and mean annual maximum surface temperatures, respectively. Point c defines the average depth of summer thaw and the top of the permafrost table. These three points vary somewhat from year to year, but when long-term running means are used to construct the plot they are reasonally stable reference points. Point d is the lowest depth of zero annual temperature fluctuation. Below this depth the temperature decreases with increasing depth. This is characteristic of the terrestrial north Polar region; the significance of this will become apparent when Fig. 2 is examined.

Figure 2 depicts the thermal regime of terrestrial permafrost at depth. The situation shown in Fig. 2 is that currently found at Prudhoe Bay, Alaska. Point e is the mean annual temperature at the depth of zero annual temperature fluctuation; it is identical with point d in fig. 1. From point e to point f, the negative temperature gradient declines and becomes positive and linear from point f to point g. This is the result of the imposition of higher surface temperatures accompanying the major climatic warming that has caused the disappearance of the continental ice sheets that in earlier times covered much of the earths land surface. By extrapolating the linear portion upward, an earlier mean annual temperature of about $-12°C$ can be deduced for the depth of zero annual temperature fluctuations. From point G downward the larger positive temperature gradient with increasing depth is due to the difference in thermal conductance between the frozen permafrost above and the unfrozen ground below. Although different with respect to the

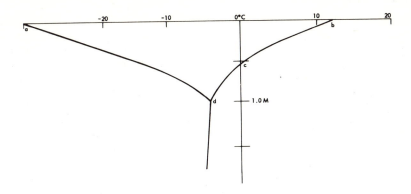

Fig. 1.   Thermal regime of permafrost, near surface
(temperature vs. depth).

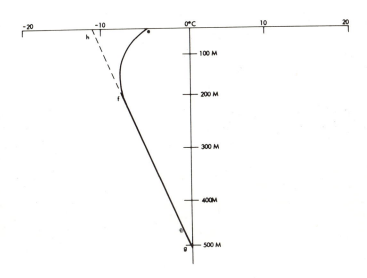

Fig. 2.   Thermal regime of permafrost, at depth (temperature vs. depth).

temperatures associated with the various points (they are some
50°C lower) the situation on Mars must be qualitatively similar.

    Viking 1 landed in the Chryse Planitia region of Mars (22.
5°N, 48.0°W). On July 20, 1976, the lander cameras revealed a
rocky undulating surface. Windblown materials were seen to be

deposited here and there in low dunes (Mutch *et al.*, 1976). A subsurface sample of windblown material (from 4 to 6 cm depth) was acquired on July 29, 1976. This sample was not analyzed immediately because of a failure to receive an "oven full" indication. On August 9, 1976, however, this sample was analyzed by the gas chromatograph/mass spectrometer. On August 21, 1976, another sample of coarse, cohesionless, granular material was acquired from the surface about 3 m away from the first sampling site (Shorthill *et al.*, 1976a). The results of the water analysis are summarized in Table I. The data show that most of the

TABLE I. *Mars Surface Water Content (%) at Chryse Planitia: the Viking GCMS Determination*[a].

	T_1	T_1	T_2
	200°C	350°C	500°C
Sample 1	<<0.1	--	0.1 to 1.0
Sample 2	--	0.1 to 1.0	<0.1 to 1.0

[a]*Samples 1 and 2 headed to temperatures T_1 and T_2. From Biemann et al. (1976).*

water is evolved during heating in the temperature range from 200 to 350°C. Additional water, however, was evolved from 350 to 500°C. This behavior is indicative of the presence of mineral hydrates. Suggestions as to the specific minerals involved are purely speculative but geothite likely is among them. The presence of clay minerals is not indicated but neither are they ruled out. For as becomes evident on detailed examination, the possibility of complete degassing of clay-adsorbed water between the time the sample was acquired and its subsequent analysis could prevent their detection. That this is the case may be deduced from studying the low-temperature adsorbtion isotherms for montmorillonite presented earlier (Anderson *et al.*, 1967). These show that at the temperature prevailing at this landing site (-31 to -187°C) the Martian regolith could retain only a monolayer of physically adsorbed water; additional free water would be in the thermodynamically more stable ice phase. This material when brought into the warm lander would be brought, in effect, into an extremely low relative humidity (10^{-3} to 10^{-5}%) and consequently, as shown in Fig. 2 (Anderson *et al.*, 1967).

Even though the Viking Lander I failed to detect the presence of physically adsorbed or free water, its demonstrated presence in the Martian atmosphere (Briggs, 1976) and the presence of mineral hydrates in the regolith is a certain indication of its presence from time to time. Conditions must be nearly always extremely dry, however. Taking the maximum values given by Farmer et al., (1976) (10 μm precipitable water in the atmospheric column) one calculates a maximum of 1×10^3 gm of water/cm^2 over the Martian surface. Distributing this quantity in a monolayer over the surfaces of the mineral grains and assuming a bulk density of the Martian regolith of 1.8 gm/cm^3 (Shorthill et al., 1976b) the regolith depth required for total, diurnal exchange of this atmospheric water is computed to be only 1.87 cm, for regolith materials having a specific surface area of 1 m^2/gm. For regolith material of 10 m^2/gm the depth required would be 0.187 cm and correspondingly less for finer and finer grained materials.

Thus it appears that Mars is a cold, dry permafrost planet. The normal form of free water is ice. Liquid water must be rare. Its most normal occurrence must be in capillaries and interfacial layers (Anderson and Morgenstern, 1973). While certainly not a favorable environment of living organisms, it is conceivable that Martian organisms, or terrestrial organisms transported to Mars, accidentally or otherwise, might occupy this environmental niche.

REFERENCES

Anderson, D. M., and Banin, A. (1975). *Origins of Life 6*, 23–26.
Anderson, D. M., and Morgenstern, N. R. (1973) (1973). Permafrost: The North American Contribution to the Second International Conference, pp. 257–288.
Anderson, D. M., Gaffney, E. S., and Low, P. F. (1967). *Science 155*, 319–322.
Biemann, K., Oro, J., Toulmin, P. III, Orgel, L. E., Nier, A. O., Anderson, D. M., Simmonds, P. G., Flory, D., Diaz, A. V., Rushneck, D. R., and Biller, J. A. (1976). *Science 194*, 72–76.
Briggs, J. (1976). Colloquium on Water in Planetary Regoliths, sponsored by NASA.
Farmer, C. B., Davies, D. W., and LaPorte, D. D., (1976). *Science 193*, 776–780.

Mutch, T. A., Binder, A. B., Huck, F. O., Levinthal, E. C.,
    Liebes, S., Jr., Morris, E. C., Patterson, W. R., Pollack,
    J. B., Sagan, C., and Taylor, G. R. (1976). *Science 193*,
    791-801.
Shorthill, R. W., Morre, H. J. II, Scott, R. F., Hutton, R. E.,
    Liebes, S., Jr., and Spitzer, C. R. (1976a). *Science 194*,
    91-9 7.
Shorthill, R. W., Hutton, R. E., Moore, H. J. II, Scott, R. F.
    and Spitzer, C. R. (1976b). *Science 193*, 805-809.
Soderblom, L. A., Malin, M. C., Cutts, J. A., and Murray, B. C.
    (1973). *J. Geophys. Res. 78*, 4197-4210.

14

AN OUTLINE OF BIOLOGICAL EVOLUTION BASED ON MACROMOLECULAR SEQUENCES[1]

Robert M. Schwartz, Margaret O. Dayhoff

National Biomedical Research Foundation
Georgetown University Medical Center, Washington, D.C.

The evolutionary origins and temporal order of the major biochemical adaptations as well as the environmental conditions under which they developed on earth should provide a guide to our understanding if life is detected on other earthlike planets. Some proteins and nucleic acids are sufficiently ancient and are changing so slowly that the record of accepted point mutations reflected in their sequences allows us to trace their evolution back close to the origin of life. Among these are the sequences of ferredoxin, 5S ribosomal RNA, and c-type cytochromes. The information in these three evolutionary trees can be combined to yield the order of divergence of the major biological groups from the most primitive bacteria to animals, including representatives of all three families of photosynthetic bacteria, numerous anerobic and aerobic bacteria, and the blue-green algae, as well as the eukaryote host and the mitochondrion and chloroplasts.

[1]*This work was supported by NASA contract NASW 2848 and NIH grants GM-08710 and RR05681.*

INTRODUCTION

Many proteins and nucleic acids are "living fossils" in the sense that their structures have been dynamically conserved by evolution over billions of years (Bryson and Vogel, 1965; Dayhoff and Eck, 1968). It is very possible that their sequences still carry sufficient information to permit us to unravel the early evolution of the biological species and the chemical processes that constitute the living world. It is now clear that many protein and nucleic acid sequences occur in recognizably related forms in eukaryotes and prokaryotes (Dayhoff, 1976a). They are believed to be of common evolutionary origin, having evolved by a great number of small changes in sequence.

The two principal computer methods for objective treatment of sequence data to elucidate evolutionary history were first described and applied more than ten years ago (Eck and Dayhoff, 1966b; Fitch and Margoliash, 1967). A vertebrate phylogeny consistent with the fossil record and with morphological considerations has been worked out independently from each of a number of proteins using such methods (McLaughlin and Dayhoff, 1973; Fitch, 1976; Goodman et al., 1974; Dayhoff, 1972). Only recently has enough sequence information become available from diverse bacterial and blue-green algal types and from the cytoplasm and organelles of eukaryotes to permit the construction of a biologically comprehensive evolutionary tree. In this chapter we shall present for the first time an evolutionary tree that extends back to the origin of the separate biological kingdoms now extant, derived by objective methods from sequence data.

Knowing the evolutionary relationships of all organisms has great predictive advantage in many areas of biology because all systems within the organisms will show a high degree of correlation with such a schema. The relative order of events in the evolution of the various metabolic pathways and their protein constituents and in the divergences of bacterial types is important in clarifying the biochemical nature of each species. The long-standing question of whether or not eukaryote organelles were originally symbiotic prokaryotes can be resolved. As a basis for comparison, such an evolutionary tree would be invaluable if life is found on other planets.

RECONSTRUCTING EVOLUTION ON THE BASIS OF SEQUENCE DATA

Evolutionary history is conveniently represented by a tree. Each point on the tree corresponds to a time, a sequence, and a species within which the sequence occurred. There is one point of earliest time corresponding to the ancestral sequence and organism. Time advances on all branches of the tree emanating from this point. The topology of the branches gives the relative order of events. During evolution, sequences in different species have gradually and independently accumulated changes yielding the sequences found today in the various organisms represented at the ends of the branches. On the trees in this chapter we have drawn the lengths of the branches proportional to the amount of evolutionary change.

In order to construct a phylogenetic tree on the basis of proteins or nucleic acids, we assume that each residue in a sequence can be treated as an inherited biological trait. This assumption implies our ability to align sequences so that changes reflect substitutions of one residue for another during the course of evolution. The accuracy with which we can deduce evolutionary connections is limited by our ability to deal with the insertions and deletions of genetic material and to correct for superimposed and parallel mutations.

Two major methods for constructing phylogenetic trees based on sequences are currently used: one proceeds by generating ancestral sequences (Dayhoff et al., 1972a; Fitch and Farris, 1974), the other by producing a least-squares fit to a matrix of evolutionary distances between the sequences (Dayhoff, 1976a; Fitch and Margoliash, 1967). The ancestral sequence method is a problem in double-minimization of inferred changes. For each possible configuration of the evolutionary tree, a set of ancestral sequences is determined that minimizes the number of inferred changes. Of these configurations, the one that minimizes the total number of changes between ancestral and known sequences is selected as the best representation of the evolutionary tree. This method also yields a set of ancestral sequences corresponding to the branch points of the tree.

In the matrix method, which was used here, we begin by calculating a matrix of percentage differences between sequences in an overall alignment; we then correct this matrix to evolutionary distances in terms of accepted point mutations per 100 residues. Two types of correction are necessary. First, we correct for inferred parallel and superimposed mutations using a scale based on our computer simulations (Dayhoff, 1976b). Second, although a small number of deleted or inserted residues in a sequence can be counted as single residue changes, large unmatched internal regions as well as ends of sequences cannot, and they are omitted from our calculations. For a given matrix, the determination of the best tree is a problem in double-minimization. For each possible configuration, a set of branch lengths is determined that provides a weighted least-squares fit between the distances given by the reconstructed matrix and those of the original matrix. The configuration that has the minimal total branch length is selected as the best solution.

To test the accuracy of both of these methods, we simulated
a number of evolutionarily related families of sequences for
which there were amino acid replacements but no insertions or
deletions. The results (Fig. 1) show that for sparse trees of
distantly related sequences the matrix method is clearly superior to the ancestral sequence method. Because this is precisely the type of tree with which we are concerned, all of the
individual trees shown here were constructed by the matrix
method.

Three superfamilies include sequences from several prokaryotes and eukaryotes: ferredoxin, 5S ribosomal RNA, and
c-type cytochromes. They are used here to construct evolutionary trees. None of these conveys an overall picture of the
course of evolution during the early Precambrian. Fortunately,
the groups of organisms and eukaryote organelles from which
sequences are available overlap in such a way that the trees
can be correlated and a composite tree can be constructed.
From this composite tree, a clear idea of the major developments
in early biological evolution and their relationships to one
another begins to emerge.

FERREDOXINS

The ferredoxins are small, iron-containing proteins that
are found in a broad spectrum of organisms and participate in
fundamental biochemical processes such as photosynthesis, nitrogen fixation, sulfate reduction, and other oxidation-reduc-
tion reactions. The tree (Fig. 2) derived from the known sequences (Barker *et al.*, 1976; Hase *et al.*, 1976b;
Tanaka *et al.*, 1974, 1975; Wada *et al.*, 1975; Hase *et al.*, 1976a)
provides a framework for the events outlined in the other evolutionary trees presented here; moreover, a gene doubling shared
by all the ferredoxin sequences makes it possible to locate the
point of earliest time in these trees. The clostridial-type
ferre-doxins, in particular, are still very similar in sequence
to the extremely ancient protein that duplicated. Most of them
are composed of fewer than the 20 common amino acids, lacking
those amino acids that are thermodynamically less stable, such as
tryptophan and histidine (Eck and Dayhoff, 1966a). From an
alignment of the first and second halves of these sequences, we
inferred an ancestral half-chain sequence. This sequence,
doubled, was included in the computations of the evolutionary
tree. Because all of the ferredoxin sequences show some evidence of gene doubling, this event must have occurred prior to
the species divergences shown here, and the ancestral sequence
must therefore be connected to the tree at the point of earliest
time. The organisms near the base of the tree, *Clostridium*,

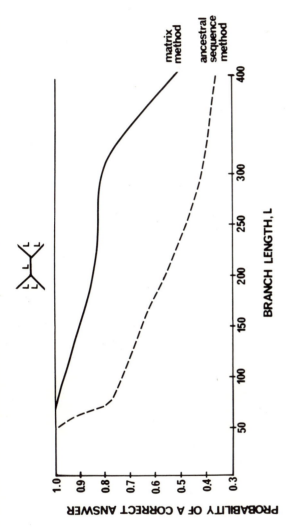

FIGURE 1. Probability of inferring the correct topology from sequences of simulated evolutionary connection. A history of five equal intervals of mutational distance was used, as shown at the top of the figure. Ten sets of sequences of 100 residues were generated for each mutational distance (L = 25, 50, 75, 100, 125, 150, 175, 200, 225, 250, 300, 400). Random events were assigned according to the average mutability and mutation pattern of each amino acid (Dayhoff et al., 1972b). In totaling the number of correct topologies inferred, a unique right answer counted 1, a unique wrong answer 0, a two-way tie 0.5, and a three-way tie 0.33. Smoothed curves through the data points are shown. The matrix method is clearly superior.

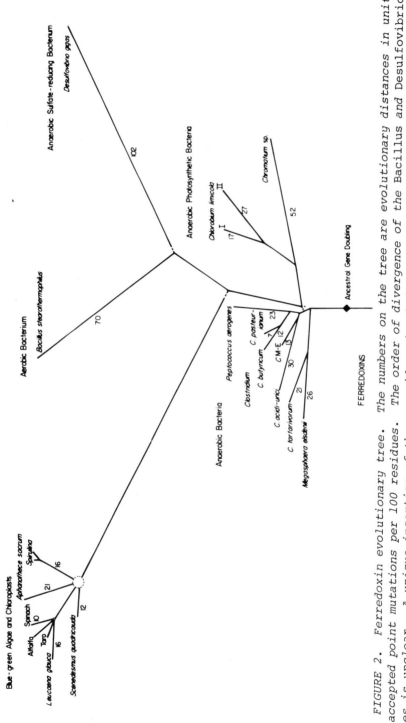

FIGURE 2. Ferredoxin evolutionary tree. The numbers on the tree are evolutionary distances in units of accepted point mutations per 100 residues. The order of divergence of the Bacillus and Desulfovibrio lines is unclear. A unique insertion of three residues in their sequences is consistent with the topology shown, but the topology with the Desulfovibrio branch coming directly off the ancestral line to the anaerobic bacteria does have nearly as short an overall length. The Spirulina species shown are S. maxima and S. platensis.

Megasphaera, and *Peptococcus*, are all anaerobic, chemoorgano-
trophic bacteria. Most species of these groups lack heme-con-
taining proteins, such as the cytochromes and catalase. *Chlorobium* and *Chromatium* diverged very early from these bacteria,
although the exact point of divergence is not clearly resolved.
Chromatium and *Chlorobium* are anaerobic bacteria capable of
photosynthesis in the presence of H_2S. Of the anaerobic bacteria shown, only *Chlorobium limicola* cannot live fermentatively,
and it is reasonable to suppose that the ability to do so is
primitive and was lost by this bacterium. The two ferredoxins
in *Chlorobium* are the result of a gene duplication within this
line.

The *Bacillus* and *Desulfovibrio* lines diverged next from the
line leading to the blue-green algae. *Bacillus* lives fermentatively under anaerobic conditions but can respire aerobically.
Desulfovibrio is a sulfate-reducing bacterium; it respires
anaerobically using sulfate as the terminal electron acceptor.
The use of sulfate by *Desulfovibrio* contrasts sharply with the
use of oxygen as the terminal electron acceptor of respiration
by *Bacillus* as well as by all of the more highly evolved forms
on this tree. This difference suggests that the divergence of
these bacteria occurred after some elements in the respiratory
chain had developed. The topology pictured here indicates that
the final elements in the respiratory chain evolved separately in
the *Bacillus, Desulfovibrio,* and blue-green algal lines.

The plant-type ferredoxins are all very closely related to
one another. The ferredoxins from the green alga *Scenedesmus*
and the higher plants are found in the chloroplasts of these
organisms. *Spirulina and Aphanothece* are representatives from
the two major divisions of blue-green algae, the filamentous
and coccoid lines (Desikachary, 1973). The topology and branch
lengths in this portion of the tree could depict the radiation
of blue-green algae followed by a symbiosis of one with the
ancestor of the higher plants. There were then subsequent
divergences in the higher plant and *Spirulina* lines.

ORIGINS OF EUKARYOTE ORGANELLES

There are two schools of thought concerning the origin of
eukaryote organelles: one is that they arose by the compart-
mentalization of the DNA within the cytoplasm of an evolving
protoeukaryote (Raff and Mahler, 1972; Uzzell and Spolsky,
1974); they arose from free-living forms that invaded a host
cell and established a symbiotic relationship with it (Margulis,
1970, 1977). In the former theory, all genes arose within the
organism; homologs found in both the nucleus and the organelles
arose by gene duplication. This theory would place the animals
and fungi together with the higher plants in the upper portion
of the ferredoxin tree subsequent to the divergence of the

blue-green algae. The eukaryotes would have developed from the
reorganization of the nuclear material of a blue-green algal
ancestor with relatively minor changes in protein and nucleic
acid sequences, if these ferredoxins are a guide. The symbiotic theory, on the other hand, proposes that the chloroplast
was originally a free-living blue-green alga; other symbionts
include the mitochondrion, which was originally a free-living
aerobic bacterium, and the flagellum and mitotic appartus, which
were derived from spirochetes. These prokaryotes separately
invaded a protoeukaryote host cell and developed a symbiotic
relationship with it. Their current status as organelles gradually evolved from these symbioses. In this theory, mitochondrial and chloroplast genes are expected to show evidence of
recent common ancestry with the separate types of contemporary
free-living prokaryote forms. The host and organelles would
occur on different branches that could also contain free-living
forms. Unfortunately, the ferredoxin tree by itself does not
permit us to distinguish between these theories because there
are no eukaryote cytoplasmic or mitochondrial sequences on this
tree.

5S RIBOSOMAL RNA

    The 5S ribosomal RNA molecule is a low molecular weight RNA
about 120 nucleotides long. It is associated with the larger
ribosomal subunit and is thought to function in the nonspecific
binding of transfer RNA to the ribosome during protein synthesis
(Monier, 1974). Because this function is independent of the
kind of amino acid, this type of molecule could be extremely
ancient, predating the contemporary form of genetic code. Sequences of 5S ribosomal RNA are known from a wide variety of
sources (Schwartz and Dayhoff, 1976; Woese *et al.*, 1975; Pribula *et al.*, 1976; Raue *et al.* 1975; Benhamou and Jordan,
1976), including aerobic and anaerobic bacteria, blue-green
algae, and the cytoplasm of several eukaryotes. The cytoplasmic
sequences, in particular, present the possibility that an
evolutionary tree based on this molecule will provide further
insight into the origin of the eukaryotes.
    We derived an evolutionary tree (Fig. 3) on the basis of
these sequences and placed its origin on the branch to the anaerobic bacterium *Clostridium*.in conformance with the ferredoxin
tree. The eukaryote sequences on this tree were all derived
from cytoplasmic ribosomes and thus, in the symbiotic theory,
describe the evolutionary history of the eukaryote host. The
branch leading to these sequences diverges from the prokaryotes
quite close to the *Bacillus* branch. Both *Bacillus* and *Esche-*

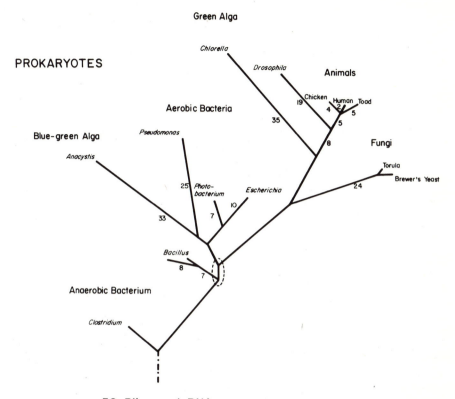

EUKARYOTES

Green Alga

Chlorella

PROKARYOTES

Drosophila

Animals

Aerobic Bacteria

Chicken Human Toad

19   4   2   5

35   5

Blue-green Alga

Pseudomonas

8   **Fungi**

Anacystis

Torula

25 *Photo-*
    bacterium   *Escherichia*

Brewer's Yeast

10   24

33   7

Bacillus

8

7

Anaerobic Bacterium

Clostridium

5S Ribosomal RNA

FIGURE 3.   *5S ribosomal RNA evolutionary tree.*   The order of *divergence for the branches leading to the eukaryotes and to* Bacillus *is not clearly resolved; the tree whose topology re- verses the order of these branches has nearly as short an over- all length.   The prokaryote species shown are* Clostridium pas- teurianum, Bacillus megaterium, B. *licheniformis,* E. coli, Pseudomonas fluorescens, *and* Anacystis nidulans.   *The* Chlorella *species is* C. pyrenoidosa.

richia are facultative aerobes, having metabolisms that are both fermentative and respiratory. The ancestral eukaryote host cell probably also had this bimodal metabolic capacity.

The two organisms that possess oxygen-evolving photosynthesis, the green alga *Chlorella* and the blue-green alga *Anacystis,* appear on opposite sides of this tree. *Anacystis* is grouped in the same family with *Aphanothece* (Desikachary, 1973), which appeared on the ferredoxin tree. These coccoid blue-green algae are certainly more closely related than the blue-green algal orders represented by *Aphanothece* and *Spirulina*. Thus, we would predict that *Aphanothece* would be found to diverge near the end of the *Anacystis* branch, preceded slightly by the divergence of the chloroplast branches. The very separate history of the cytoplasmic sequence points to a symbiotic origin of the chloroplasts.

C-TYPE CYTOCHROMES

The evolutionary tree (Fig. 4) based on c-type cytochrome sequences (Dayhoff and Barker, 1976; Ambler *et al.,* 1976; Aitken, 1976) is important to an understanding of the origin and evolution of the mitochondrion. Cytochrome c is coded in the nucleus but functions in the mitochondrion. This is usually explained in the symbiotic theory by transfer of genetic information, including the cytochrome c gene, from the invading aerobic bacterium to the protoeukaryote host during the development of their current relationship. Such a transfer seems reasonable because the energy-converting function of the mitochondrion is so essential to eukaryotes that this symbiosis must have occurred and been stabilized very early in their evolution. In the nonsymbiotic theory, genetic rearrangement is also an essential feature.

This tree places the eukaryote mitochondrion in the portion of the tree with aerobic bacteria after their divergence from the blue-green algae and chloroplasts. The cytochrome c sequences are on a branch that most recently diverged from cytochrome c_2 of the nonsulfur, purple, photosynthetic bacteria; together these diverged from the branch leading to cytochrome c_{551} from strict aerobes such as *Pseudomonas*. This contrasts with the evolution of the eukaryote host depicted by the 5S ribosomal RNA tree. There the eukaryote host diverged among the facultatively aerobic bacteria, such as *Bacillus,* from the line leading toward the blue-green algae. *Pseudomonas* diverged from this line somewhat later. The topology of the animal subtree derived from cytochrome c (McLaughlin and Dayhoff, 1973), reflecting the evolutionary history of the mitochondrion, is consistent with that of the eukaryote host derived from 5S ribosomal RNA because the divergences are subsequent to the mitochondrial invasion.

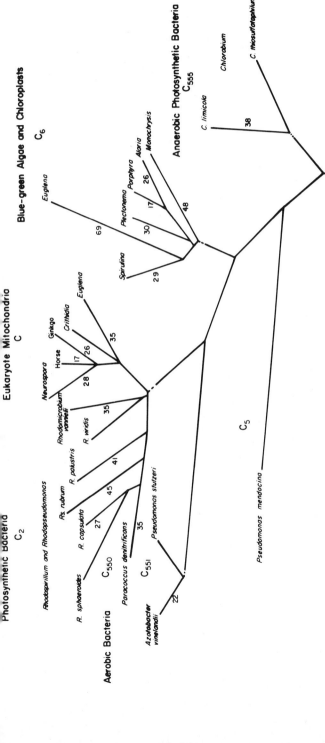

FIGURE 4. *c-Type cytochrome evolutionary tree. Cytochrome c_{550} and all of the c_2 are separated from the other c-type cytochromes by a unique deletion close to the heme-binding cysteine in their sequences and on this basis were placed on a separate branch from the cytochrome c sequences. Euglena and Crithidia have been placed on a single branch together, a configuration slightly less than optimal, because they share a unique mutation of the active cysteine. The connections of the c_{551} and c_{555} sequences have been centered. The genera Rhodospirillum and Rhodopseudomonas are abbreviated Rs. and R., respectively. Cytochrome c_6 sequences were taken from the following species: Spirulina maxima, Monochrysis lutheri, Porphyra tenera, Euglena gracilis, Alaria esculenta, and Plectonema boryanum; cytochrome c sequences were taken from the protists Euglena gracilis and Crithidia oncopelti. Chlorobium limicola and C. thiosulfatophilum have been reidentified as Prosthecochloris aestuarii and C. limicola, respectively.*

Cytochrome c_2 is found in bacteria such as *Rhodomicrobium* and *Rhodopseudomonas*. These bacteria are photosynthetic anaerobically but respire aerobically. *Paracoccus denitrificans* possesses cytochrome c_{550}, which is very similar to cytochrome c_2 along its entire length. *Paracoccus*, like the mitochondrion, has apparently lost its photosynthetic ability.

Two different c-type cytochromes, c_5 and c_{551}, are found in *Pseudomonas*, a nonphotosynthetic bacterium. It is probable that these are the result of a gene duplication early in the tree with subsequent loss of the c_5 gene in some lines.

Cytochrome c_6 is found in the photosynthetic lamellae of blue-green algae and the chloroplasts of eukaryotes, where it functions in the electron transport chain between photosystems I and II. As in the ferredoxin tree, there is a close similarity between the blue-green alga *Spirulina* and the various eukaryote algal chloroplasts; as in the 5S ribosomal RNA tree, the blue-green algae are most closely related to strictly aerobic bacteria, such as *Pseudomonas*. The topology and branch lengths reflect a symbiotic origin for photosynthesis in eukaryotes. It is not possible to locate precisely the point at which the main tree connects to the c_6 subtree. However, there are separate branches leading to the two filamentous blue-green algae, *Spirulina* and *Plectonema*, and therefore this subtree must reflect, at least in part, the evolutionary relationships among invading blue-green algae rather than the speciation of eukaryotes; some of the eukaryote algal chloroplasts appear to be derived from different symbiotic associations.

The point of earliest time on the c-type cytochrome tree was placed near *Chlorobium*, an anaerobic, obligate photosynthetic bacterium. This is consistent with the position of the sequence from *Chlorobium* in the ferredoxin tree.

COMPOSITE EVOLUTIONARY TREE

Each of the individual trees we have presented contains information about the early course of biological evolution. By comparing these trees we have developed information that a composite evolutionary tree would contain; for example, these comparisons support symbiotic origins for the eukaryote organelles. In order to obtain the overview that is inherent in the individual trees, we used their topologies and evolutionary distances to construct a composite tree (Fig. 5).

The composite evolutionary tree was constructed from a composite matrix using the same methods that we used to construct the individual trees. This matrix included the six species that appear on at least two of the three individual trees

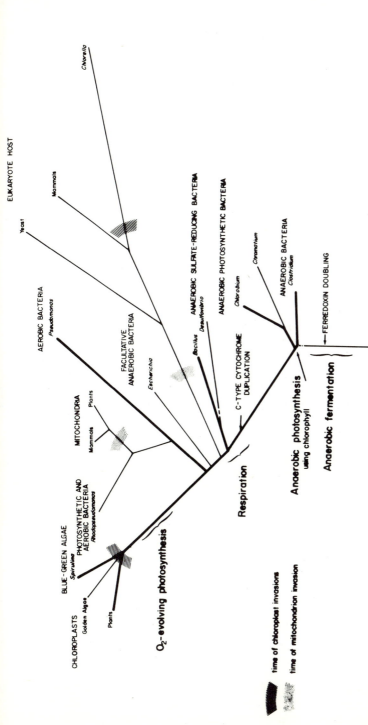

FIGURE 5. Composite evolutionary tree. This tree presents an overview of early evolution based on ferredoxin, c-type cytochrome, and 5S ribosomal RNA sequences. The heavy lines represent a tree calculated from a matrix of evolutionary distances combining the individual trees. The lighter lines represent branches scaled from individual trees and added to the combined tree. The point in evolution at which the mitochondrial invasion occurred is stippled; the chloroplast invasion is shaded.

(their branches are represented by the heavy lines in Fig. 5).
First, the trees were scaled so that distances were comparable.
A combined matrix of distances between the six species was
calculated by averaging these scaled contributions. As previously, for each possible configuration of the combined tree,
a set of branch lengths was determined that provided a weighted
least-squares fit between the matrix of distances between
species and a matrix reconstructed from the tree. The configuration with the shortest overall branch length was chosen.
This configuration is the one that is also consistent with all
of the individual trees. Finally, branches found in only one
tree were scaled in length and added to the composite tree,
maintaining the relative internodal distances (these are represented by light lines in Fig. 5).

The genetic doubling of the clostridial-type ferredoxins
allows us to place the point of earliest time in this tree.
Moreover, because the species whose sequences have changed
least since this doubling event were all anaerobic, chemoorgan-
otrophic bacteria, it is likely that the ability to live fermentatively is primitive. The composite tree describes the
evolution of photosynthesis using chlorophyll, starting with
the development of the ability to synthesize the compound. All
three families of photosynthetic bacteria are represented:
Chromatiaceae, Chlorobiaceae, and Rhodospirillaceae. The divergence of the Chromatiaceae and the Chlorobiaceae from the
other anaerobic bacteria was quite early, and it is clear that
this type of photosynthesis arose at a very early stage in
evolution and has not changed much. The Rhodospirillaceae are
aerobic heterotrophs with the ability to photosynthesize anaerobically. They are much more highly evolved than the other
photosynthetic bacteria.

On the branch leading to the more highly evolved organisms,
the first major development shown is aerobic respiration. As
we noted in discussing the ferredoxin evolutionary tree, the
divergence of *Bacillus* and *Desulfovibrio* probably marks the
appearance of some components of this adaptation. The final
elements in this adaptation evolved separately because these
groups differ in their terminal electron acceptor of respiration. Additionally, the divergence of cytochrome c_5 occurred
just prior to this time. In the c-type cytochrome tree, this
branch was puzzling; in the context of the evolution of a respiratory metabolism, it is interesting because such a duplication of a c-type cytochrome gene could have contributed to
the evolutionary plasticity required in the development of
respiration.

The eukaryote host diverged at about the same time as *Bacillus* and *Escherichia*. Both of these bacteria are facultative
anaerobes and usually live in environments with almost no oxy-

gen, such as soil and the intestines of animals. The ancestral
eukaryote host could well have been facultative at the time of
its divergence.

The bacterium that has become the mitochondrion was most
closely related to the third family of photosynthetic bacteria,
the Rhodospirillaceae. Photosynthesis was lost either before
or shortly after this ancestral bacterium invaded the host. It
seems reasonable to suppose that the aerobic respiratory metabolism of this invading protomitochondrion was more effective
than that of the host and that the host lost any primitive system that it might have had, including its cytochrome c.

The final biochemical adaptation depicted here is the development of photosystem II. The blue-green algae and chloroplasts
of the eukaryotes uniquely possess oxygen-evolving photosynthesis, and this capacity appears to have evolved only once. It
combined the new biochemical adaptation, photosystem II, with
proteins modified from two earlier adaptations, bacterial photosynthesis and respiration. Clearly, the chloroplasts, like the
mitochondria, are the result of symbiotic associations with
the eukaryote host.

The composite tree makes it particularly clear that three
of the branches that contribute to the eukaryote are distinctly
separate; each is closely related to free-living prokaryotes.
The chloroplasts share a recent ancestry with the blue-green
algae; the mitochondrion shares a recent ancestry with certain
photosynthetic bacteria, the Rhodospirillaceae; whereas the
eukaryote host diverged from the other groups at a considerably
earlier time along with Bacillus and Escherichia.

If current estimates of the antiquity of life on earth are
correct, bacteria very much like Clostridium lived more than
3.2 billion years ago. Bacterial photosynthesis evolved nearly
that long ago, and it seems reasonable, in view of our composite tree, to attribute the most ancient stromatolites, formed
about 2.8 billion years ago, to early photosynthetic bacteria.
Blue-green algae evolved later. The tree shows that by the time
oxygen-evolving photosynthesis originated in the blue-green algal
line, there must have been a great diversity of morphological
types, including ancestral bacteria from most of the major groups
pictured on the composite tree. This time probably corresponds
to the great increase in complexity of the fossil record about
2 billion years ago. Our composite tree suggests that respiration preceded oxygen-evolving photosynthesis. The latter process
was, in large measure, responsible for the final transition to
the present-day oxygen level in the atmosphere. Judging from
the relative branch lengths on the tree, the mitochondrial invasion

occurred during this transition. Finally, perhaps 1.1 billion years ago, serveral independent symbioses between protoeukaryotes and various blue-green algae resulted in photosynthetic eukaryotes; some of these developed into modern eukaryote algae, whereas a single line appears to have evolved into the higher plants. This broad outline of early events in the emergence of life can be refined as new sequence information becomes available. It is our expectation that eventually all of the biochemical components of intermediary metabolism will be correlated with the development of the prokaryote types and their metabolic capacities.

ACKNOWLEDGMENTS

We thank W. C. Barker and B. C. Orcutt for their helpful discussions and criticism and K. Lawson, who drafted the figures.

REFERENCES

Aitken, A., (1976). *Nature 263,* 793.

Ambler, R. P., Meyer, T. E., and Kamen, M. D. (1976). *Proc. Nat. Acad. Sci. USA 73,* 472.

Barker, W. C., Schwartz, R. M., and Dayhoff, M. O. (1976). In "Atlas of Protein Sequence and Structure," Vol. 5, Suppl. 2 (M. O. Dayhoff, ed.) p. 51. Nat. Biomed. Res. Found., Washington, D.C. Respiratory protein sequences are collected here. K. T. Yasunobu and H. Matsubara are responsible for elucidating many of these ferredoxin sequences.

Benhamou, J., and Jordan, B. R. (1976). *FEBS Lett. 62,* 146.

Bryson, V., and Vogel, H. J. eds. (1965). "Evolving Genes and Proteins." Academic Press, New York.

Dayhoff, M. O., ed. (1972). "Atlas of Protein Sequence and Structure," Vol. 5. Nat. Biomed. Res. Found., Washington, D. C.

Dayhoff, M. O. (1976a). *Fed. Proc. 35,* 2132.

Dayhoff, M. O., ed. (1976b). "Atlas of Protein Sequence and Structure," Vol. 5, Suppl. 2, p. 313. Nat. Biomed. Res. Found., Washington, D.C.

Dayhoff, M. O., and Barker, W. C. (1976). In "Atlas of Protein Sequence and Structure," Vol. 5, Suppl. 2 (M. O. Dayhoff, ed.), p. 25. Nat. Biomed. Res. Found., Washington, D.C. C-type cytochrome sequences are collected here. R. Ambler, E. Margoliash, D. Boulter, E. Smith, M. Kamen, and G. Pettigrew are responsible for many of these sequences.

Dayhoff, M. O., and Eck, R. V., eds. (1968). "Atlas of Protein
    Sequence and Structure," Vol. 3. Nat. Biomed. Res. Found.,
    Washington, D.C.
Dayhoff, M. O., Park, C. M., and McLaughlin, P. J. (1972a).
    In "Atlas of Protein Sequence and Structure," Vol. 5,
    (M. O. Dayhoff, ed.), p. 7. Nat. Biomed. Res. Found.,
    Washington, D. C.
Dayhoff, M. O., Eck, R. V., and Park, C. M. (1972b), In "Atlas
    of Protein Sequence and Structure," Vol. 5 (M. O. Dayhoff,
    ed.), p. 89. Nat. Biomed. Res. Found., Washington, D.C.
Desikachary, T. V. In "The Biology of Blue-Green Algae" (N. G.
    Carr and B. A. Whitton, eds.), p. 473. Univ. of California
    Press, Berkeley and Los Angeles.
Eck, R. V., and Dayhoff, M. O. (1966a). *Science 152,* 263.
Eck, R. V., and Dayhoff, M. O. (1966b). "Atlas of Protein
    Sequence and Structure," Vol. 2. Nat. Biomed. Res. Found.,
    Washington, D. C.
Fitch, W. M. (1976). *J. Mol. Evol. 8,* 13.
Fitch, W. M., and Farris, J. S. (1974).   *J. Mol. Evol. 3,* 263.
Fitch, W. M., and Margoliash, E. (1967). *Science 155,* 279.
Goodman, M.,   Moore, G. W., Barnabas, J., and Matsuda, G.
    (1974). *J. Mol. Evol. 3,* 1.
Hase, T., Ohmiya, N., Matsubara, H., Mullinger, R. N., Rao, K.
    K., and Hall, D. O. (1976a). *Biochem. J. 159,* 55.
Hase, T., Wada, K., and Matsubara, H. (1976b). *J. Biochem. 79,*
    329.
McLaughlin, P. J., and Dayhoff, M. O. (1973). *J. Mol. Evol. 2,*
    99.
Margulis, L. (1970). "Origin of Eukaryotic Cells." Yale Univ.
    Press, New Haven, Conn.
Margulis, L. (1977). In "Handbook of Genetics," Vol.1 (R. C.
    King, ed.). Plenum, New York.
Monier, R. (1974). In "Ribosomes" (M. Nomura, A. Tissieres, and P.
    Lengyel, eds.), p. 141. Cold Spring Harbor Laboratory, Cold
    Spring Harbor, New York.
Pribula, C. D., Fox, G. E., and Woese, C. R. (1976). *FEBS Lett.
    64,* 350.
Raff, R. A., and Mahler, H. R. (1972). *Science 177,* 575.
Raue, H. A., Stoof, T. J., and Planta, R. J. (1975). *Eur. J.
    Biochem. 59,* 35.
Schwartz, R., and Dayhoff, M. O. (1976). In "Atlas of Protein
    Sequence and Structure," Vol. 5, Suppl. 2 (M. O. Dayhoff, ed.),
    p. 293. Nat. Biomed. Res. Found., Washington, D. C. 5S
    ribosomal RNA sequences are collected here; B. G. Forget and
    S. M. Weissman and their co-workers are responsible for many
    of these sequences.
Tanaka, M., Haniu, M., Yasunobu, K. T., Rao, K. K. and Hall,
    D. O., (1974). *Biochemistry 13,* 5284.

Tanaka, M., Haniu, M., Yasunobu, K. T., Rao, K. K., and Hall, D. O. (1975). *Biochemistry 14*, 1938 and 5535.

Uzzell, T., and Spolsky, C. (1974). *Am. Sci. 62*, 334.

Wada, K., Hase, T., Tokunaga, H., and Matsubara, H. (1975). *FEBS Lett. 55*, 102.

Woese, C. R., Pribula, C. D., Fox, G. E., and Zablen, L. B. (1975). *J. Mol. Evol. 5*, 35.

15

MOLECULAR FOSSILS FROM THE PRECAMBRIAN NONESUCH SHALE

Thomas C. Hoering

Geophysical Laboratory, Carnegie Institution of Washington
Washington, D.C.

Organic geochemistry has played an important role in describing the life that existed on earth before the onset of an abundant fossil record at the start of the Paleozoic era. A significant part of the work has been the discovery of ancient molecules that can be related to precursors found in living organisms (MdKirdy, 1974); these organic compounds have been called "molecular fossils." Recent advances in experimental techniques and in the understanding of the diagenesis of organic molecules over long periods of time make it possible to search for molecular fossils in Precambrian rocks with certainty.

Well-preserved organic matter as old as the Precambrian is rare because most sedimentary rocks have been extensively metamorphosed; however, the Nonesuch Shale exposed in the White Pine copper mine, White Pine, Michigan, is an exception (White and Wright, 1966). This highly carbonaceous, fine-grained shale contains mixed-layer minerals derived from clays and an aluminous serpentine. Although the upper temperature limit for the stability of these minerals is not known exactly, the evidence indicates that the rock has not been heated above 200°C. Radiometric dating of associated igneous rocks has shown that the shale is between 1.0 and 1.1 b.y. old. The Nonesuch Shale is a sedimentary copper deposit, and the ore exists as sulfide minerals and native copper. The ore is not associated with hydrothermal activity or with faulting. A petroleum seep occurs within the copper mineralization, and the rock matrix contains well-preserved kerogen. Some features of the organic matter in this shale have been studied. It is highly probable that the petroleum originated in the shale and has not migrated from younger formations (Barghoorn *et al.*,

1965; Eglinton *et al.*, 1966; Hoeirng, 1967). The petroleum is
worthy of detailed examination because it represents the remains of life dating back one-fourth of geological time. Large
quantities of it are available for study, and problems due to
contamination by younger organic matter appear to be minimal.

Saturated hydrocarbons are very stable and represent the
best classes of compounds to study for evidence of Precambrian
life. The petroleum seep in the Nonesuch Shale contains 61 wt%
saturated hydrocarbons. Figure 1 gives the molecular structure
of six of the best understood types and their biological precursors. The exact mechanism by which the precursors have
been transformed into hydrocarbons is not known; in every case,
however, a few plausible chemical reactions would be sufficient.

The mixture of molecules found in petroleum is one of
the most complex in all of chemistry. An extensive separation
is needed before pure compounds can be isolated and identified.
Many methods for fractionating saturated hydrocarbons and
identifying them are known, and two recent developments have
made the present study possible. Hydrocarbons are chemically
inert, but separation into classes based on molecular shape and
size is possible by exclusion chromatography. Tandem gas
chromography-mass spectrometry is one of the most powerful
tools available to the organic geochemist for identifying components of mixtures. In complex systems, however, the amount
of data that is collected is too large for manual handling. It
is possible, by means of computer-based data systems, to
collect and store mass spectra continuously during a gas-chro-
matographic run. To use the data efficiently, programs for re-
calling the spectra and presenting them in numerical or graph-
ical forms are employed.

Figure 2 outlines the separation scheme that was used on
the petroleum seep. A large quantity (13.7 g) of the crude oil
was collected by rinsing it from specimens of the wall rock
with benzene and subsequently evaporating the solvent. The
general stratigy of the separation is as follows. The saturated hydrocarbons are isolated by silica gel chromatography
and separated into four major classes on the basis of molecular
shape and size by inclusion into crystals of synthetic zeolite,
urea, and thiourea (adduct formation). Each of these classes
is further fractionated into six or more subgroups by exclusion
chromatography on a 3 m by 15 mm column of Sephadex LH-20 ex-
panded in benzene-methanol. The separations are monitored at
each step by gas-liquid chromatography. When necessary an
additional step of exclusion chromatography on an alumina column is made to simplify the mixtures. If the subgroups are
still too complex to give unambiguous mass spectra, individual
pure compounds are isolated by multiple, preparative gas chromatography on a series of substrates.

As the separation proceeded, it became evident that over half of the fractions were too complex to be resolved by existing techniques. The only practical course qas to focus on the fractions that would contain the six classes of hydrocarbons shown in Fig. 1.

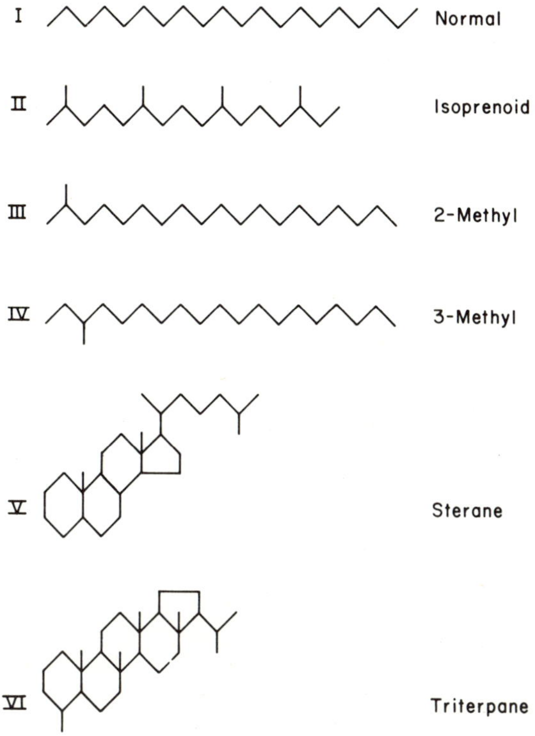

I   Normal

II   Isoprenoid

III 2-Methyl

IV   3-Methyl

V   Sterane

VI   Triterpane

FIGURE 1. *Molecular structure of six saturated hydrocarbon molecular fossils. The probable sources of these compound types are as follows. (I) Normal: fatty acids and alcohols common to most microorganisms. (II) Isoprenoid (with 20 or less carbon atoms): phytol side chain of chlorophyll. (III, IV) 2- and 3-methyl substituted: bacterial lipids. (V) Stearane: sterols of eucaryotic organisms. (VI) Triterpane: minor triterpenoid constituents of photosynthetic microorganisms.*

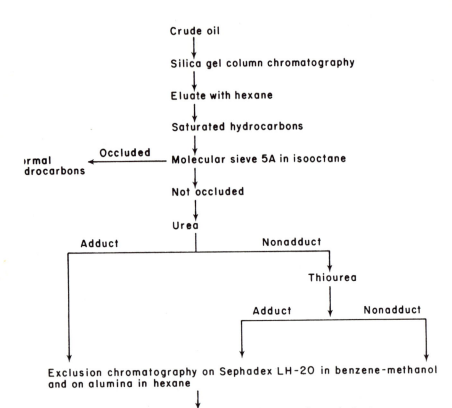

Crude oil

Silica gel column chromatography

Eluate with hexane

Saturated hydrocarbons

ormal ◄——— Occluded ——— Molecular sieve 5A in isooctane
drocarbons

Not occluded

Urea

Adduct   Nonadduct

Thiourea

Adduct   Nonadduct

Exclusion chromatography on Sephadex LH-20 in benzene-methanol
and on alumina in hexane

Preparative gas-liquid chromatography on polar substrates

Gas chromatography-mass spectrometry using a 30 m by 0.508 mm
porous layer open tube column coated with Dexil 300

*FIGURE 2.   Separation scheme used to fractionate saturated
hydrocarbons.*

Some results are illustrated by the gas chromatograms of
Figs. 3-7. Figure 3 shows a chromatogram of the total hydrocarbons isolated from the petroleum by silica gel chromatography. It is dominated by peaks due to the normal alkanes (I of
Fig. 1), which constitute 21 wt%. The normal alkanes were
easily removed by occlusion into molecular sieve 5A, a synthetic zeolite. The remaining compounds gave the exceedingly complex chromatogram shown in Fig. 4. When this mixture was exposed to urea crystallizing from a benzene-methanol solution,
27 wt% of the initial hydrocarbons were precipitated. This
adduct is known to contain linear molecules having a small
amount of branching, such as III and IV of Fig. 1. The urea

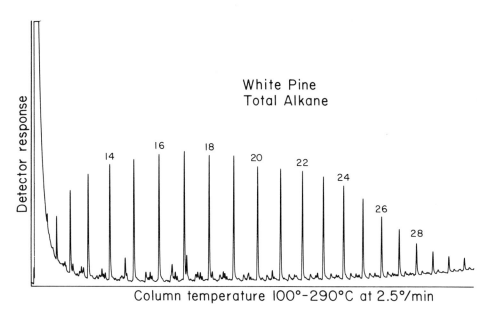

FIGURE 3. Gas chromatogram of total saturated hydrocarbon
fraction from oil seep in Nonesuch Shale at the White Pine locality. The chromatograms in Figs. 3-7 were made with a 90 m
x 0.75 mm i.d. glass, porous-layer, open-tube capillary column
coated with SE-30. A short precolumn of 0.160 mm i.d. packed
with 3% SE-30 was used. No stream splitting was employed. The
helium flow rate was 8 cm³/min. The column temperature was
programmed linearly from 100 to 290°C at a rate of 2.5°/min and
held isothermally at the upper limit for 20 min. The prominent
peaks are due to normal hydrocarbons with carbon numbers as indicated on the chromatogram.

adduct was fractionated by exclusion chromatography, and fraction 4, which amounted to 8.5 wt% of the initial hydrocarbons,
was isolated. It gave the chromatogram shown in Fig. 5. After
additional separation alumina and preparative gas chromatography of selected peaks, the identifications shown in the figure
were made.
    The remaining mixture was then exposed to thiourea crys-
tallizing from a chloroform-methanol solution, and 7.2 wt% of
the initial hydrocarbons were precipitated. This operation is
known to separate isoprenoid hydrocarbons and certain sterane
hydrocarbons (II and V of Fig. 1). The thiourea adduct was
recovered and fractionated by exclusion chromatography. Fraction 8, having 1.2 wt% of the initial hydrocarbons, was isolated. The calibration of the exclusion column with pure com-

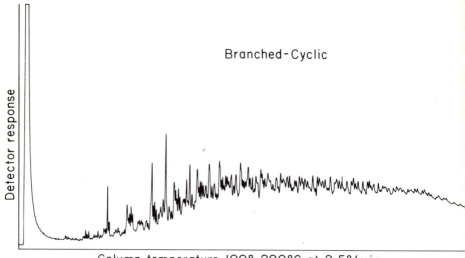

Branched-Cyclic

Detector response

Column temperature 100°-290°C at 2.5°/min

FIGURE 4.   Gas chromatogram of saturated hydrocarbons after
treatment with molecular sieve 5A to remove normal hydrocarbons.

pounds showed that both isoprenoid and sterane hydrocarbons,
if present, would be concentrated in it. The gas chromatogram
of Fig. 6 and accompanying identifications show that a homologous series of isoprenoid hydrocarbons is present, but no trace
of steranes could be detected. The identification of the iso-
prenoids was verified by the coinjection technique. Pure
synthetic isoprenoid hydrocarbons are not available for comparison. The homologous series that exists in the Green River
Shale has been studied intensively, however, and the structures of its members are known. A sample was isolated from the
Green River Shale and injected simultaneously into the gas
chromatograph with a portion of fraction 8 from the Nonesuch
petroleum seep. The isoprenoid components of the two sets
eluted at exactly the same time and gave identical mass spectra.
      The remaining hydrocarbons that did not form a stable
adduct were extremely complex and did not yield appreciably simpler fractions when separated by exclusion chromatography.
Figure 7 shows the gas chromatogram of one of the fractions.
This mixture is too complex to be resolved.
      There was no indication in any of the fractions studied of
the presence of sterane and triterpane hydrocarbons. A more
careful search was made using mass chromatography. In this
technique, the mass spectrometer is used as a very specific and

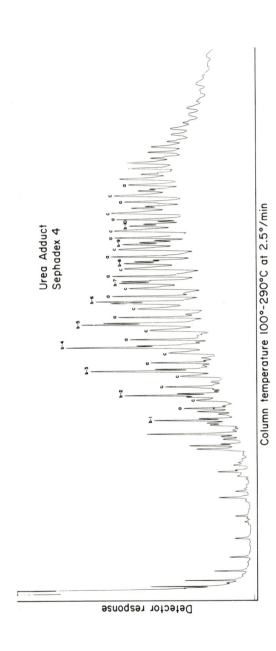

FIGURE 5. Gas chromatogram of hydrocarbons that formed an adduct with urea and were eluted in fraction 4 from the Sephadex column. Peaks labeled with (a) are due to normal hydrocarbons not completely removed by molecular sieve 5A. The labeled peaks run from C_{19} H_{40} to $C_{30}H_{62}$. The sets of two peaks designated (b-1) to (b-10) are due to 2-methyl- and 3-methyl-substituted hydrocarbons and run from $C_{19}H_{40}$ to $C_{28}H_{58}$. The broad peaks labeled (c) are due to unresolvable mixtures of monomethyl- and dimethyl-substituted linear hydrocarbons with the branching points in interior positions of the carbon chain.

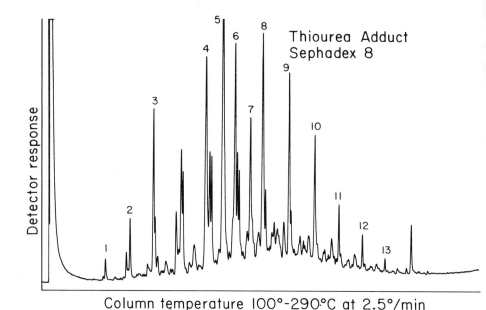

Column temperature 100°-290°C at 2.5°/min

FIGURE 6. Gas chromatogram of hydrocarbons that formed a
thiourea adduct and were eluted in fraction 8 from the Sephadex
column. Identification of peaks are as follows. Isoprenoid
hydrocarbons: (1) 2, 6, 10 trimethyl undecane; (2) 2, 6, 10
trimethyl dodecane; (farnesane); (3) 2, 6, 10 trimethyl tride-
cane; (4) 2, 6, 10 trimethyl pentadecane; (5) 2, 6, 10, 14
tetramethyl pentadecane (pristane); (7) 2, 6, 10, 14 tetramethyl
hexadecane (phytane). Alkyl cyclohexanes: (6) n-undecyl cyclohexane; (8) n-dodecyl cyclohexane; (9) n-tridecyl cyclohexane;
(10) n-tetradecyl cyclohexane; (11) n-pentadecyl cyclohexane;
(12) n-hexadecyl cyclohexane; (13) n-heptadecyl cyclohexane.

sensitive detector to monitor the output of the gas chromatographic column. The spectrometer is set to detect a specific
ion that is characteristic for a class of compounds. Ions at
mass 217 and 191 are prominent and unique features of the mass
spectra of steranes and triterpanes, respectively. Although
nanogram quantities of these compounds are detectable by this
technique, no trace of them was found in any of the fractions
at gas chromatographic retention times where they are expected
to elute. It can be concluded that they constitute less than
50 ppm of the hydrocarbons in the Nonesuch oil seep.
    Dehydrogenation by sulfur converts steroids and sterane
hydrocarbons into distinctive polynuclear aromatic hydrocarbons
called "Diels hydrocarbons." A search was made for this class
of compounds as follows. An aromatic hydrocarbon fraction was
eluted from the silica gel column (Fig. 1) with benzene after

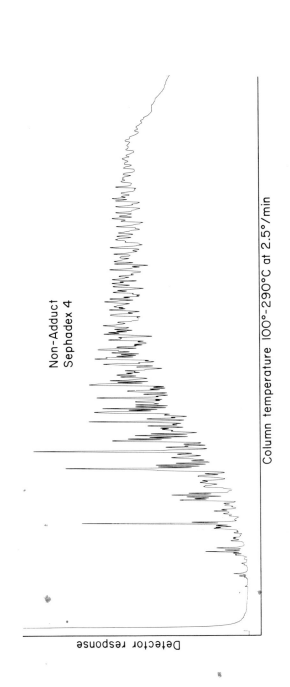

FIGURE 7. Gas chromatogram of hydrocarbons not forming a stable adduct and eluting in fraction 4 from the Sephadex column.

the saturated hydrocarbons were removed. Polynuclear aromatic hydrocarbons were separated and identified by the procedure of Giger and Blumer (1974), which was optimized for detecting Diels hydrocarbons. Synthetic standards were prepared by the reacting cholesterol with selenium (Fieser and Fieser, 1959).

The results of the search were negative. The petroleum contains only a minor amount of polynuclear aromatic hydrocarbons having a very complex composition. There was no detectable amount of Diels hydrocarbons or their isomers. Sterane hydrocarbons have not disappeared by their pathway.

The optical activity of high-boiling petroleum fractions is due, in part, to sterane and triterpane hydrocarbons. The rotation of the plane of polarization of light is caused by molecules containing a center of asymmetry. Such molecules are common in living organisms and the observation of optical activity in sedimentary organic matter is taken as proof of biological precursors.

A careful search was made for optical activity in the Precambrian hydrocarbons. Each of the Sephadex column fractions (Fig. 1) of the urea adduct, thiourea adduct, and nonadduct hydrocarbons was examined. A Bendix-Ericson Electronic Polarimeter was used at the wavelength of the sodium D line. A pair of matched cells was thermostated at 25° C. One cell was filled with the solvent (hexane) for a control and the other was filled with a solution of the hydrocarbon. Rotation was measured by alternating the solvent and sample cells. Calibration of the instrument with known, optically active compounds showed that a minimum of 0.0007° rotation was detectable. The results of the search for optical activity were negative and in disagreement with those of Barghoorn et al. (1965).

It has been shown (Hoering, 1977) that a petroleum believed to have been exposed to a high temperature during its history contained olefinic hydrocarbons. A search was made for this class of hydrocarbons in the Precambrian petroleum. Olefins were concentrated by column chromatography with silver nitrate loaded silica gel and gradient elution with hexane-diethyl ether mixtures. There was no detectable infrared absorbtion due to trans olefin bonds at 10.3 microns. The results are consistent with the low-temperature history of the Nonesuch Shale.

Petroleum and sedimentary rock extracts contain small amounts of fatty acids and other oxygenated compounds that can serve as molecular fossils. Such compounds are eluted from silica gel with methanol during the initial separation step of Fig. 2. A search for oxygenated compounds in the petroleum was made by the procedure of Hoering (1971). Briefly, the procedure consists of preparing deuterated hydrocarbons as derivatives and can be illustrated as follows, using fatty acid as an example:

$$R-CH_2-COOH \xrightarrow{\text{LiAlD}_4} R-CH_2-CD_2OH$$

$$R-CH_2-CD_2OH \xrightarrow{\text{HI}} R-CH_2-CD_2I$$

$$R-CH_2-CD_2I \xrightarrow{\text{LiAlD}_4} R-CH_2-CD_3$$

The extent and position of the deuterium labeling indicates the nature of the precursor. However, no deuterated hydrocarbons were derived from the methanol eluate of the petroleum.

The White Pine petroleum is found in well-preserved sections of the carbonaceous Nonesuch Shale. The rocks show no evidence of exposure to high temperatures. The Striped Marker Zone of the Nonesuch Shale (White and Wright, 1966) lies about 40 ft above the petroliferous copper zone and consists of a well-preserved, light-brown, calcareous mudstone. The insoluble organic matter (kerogen) was isolated from the mudstone by demineralizing it with hydrochloric and hydrofluoric acid. The resulting kerogen concentrate was then oxidized briefly with hot chromic acid in 3 m sulfuric acid. Small amounts of fatty acids (approximately 100 ppm of the starting carbon) could be extracted from the reaction mixture. The acids showed a smooth distribution of molecules with odd and even numbers of carbon atoms. Another portion of the kerogen concentrate was subjected to destructive hydrogenolysis (Hunt and Forsman, 1958) at 380°C and 2000 psi hydrogen. A small yield of saturated hydrocarbons (about 30 ppm of the starting carbon) was obtained. The experimental results indicate that detectable amounts of recognizable organic structures persist in the shale. Although not conclusive, the Nonesuch Shale is a possible source rock for the petroleum.

The molecular fossil concept is a straightforward and powerful method for studying paleobiochemical processes. Organic molecules can exist in many isomers, but biological processes are very selective and only a few of the possible isomers are actually used. On the other hand, diagenetic alteration of organic matter over long periods of time is relatively nonselective and produces a complex suite of isomers.

For example, there are over 2 million possible isomers with the formula $C_{20}H_{42}$, yet only four of them (normal, isoprenoid, 2-methyl, and 3-methyl) have an appreciable concentration. These compounds are all molecular fossils. Other fractions are very complex and seem to have been diagenetically altered. The fossil record has been obscured.

The presence of an abundant amount of normal hydrocarbons indicates that lipid synthesis was prevalent a billion years ago and had many of the same properties it has had subsequently.

Isoprenoid hydrocarbons with less than 20 carbon atoms point
clearly to the widespread use of chlorophyll as a photosynthetic pigment. The preponderance of 2- and 3-methyl-substituted
hydrocarbons in the urea adduct over all of the other possible
positional isomers suggests, but does not prove, a large contribution of bacterial lipids. The uncertainty arises because
the mechanism of the diagenesis of organic matter is not well
understood. Although it is believed that nonbiological transformations would not preferentially produce this pair of isomers, the hypothesis is not well documented.

There are limitations at present on the use of hydrocarbons
as molecular fossils. Even though an extensive separation
scheme was used in this study, only a minor fraction of the hydrocarbons could be identified. The mixtures were extremely
complex, and pure synthetic standard compounds are lacking.
Some major classes of compounds were identified (the internally
branched hydrocarbons in Fig 6 and the cycloalkanes in Fig 7)
but could not be related to known biological precursors.
Possibly their source lies in nonbiological transformation during diagenesis.

A number of molecular fossils are well preserved in the Precambrian petroleum, but the absence of detectable sterane hydrocarbons is puzzling. Such compounds are fairly common in
younger rocks and have been derived from sterols, known to exist in unicellular organisms. Sterols are interesting compounds for evolutionary studies in the Precambrian. They require
molecular oxygen for their synthesis and are used primarily by
nucleated cells. Since it is generally agreed that the earth's
atmosphere was oxygenated, at least partially, by 1100 million
years ago, there are several possibilities for the lack of
steranes. Either sterols had not been invented by that time
or else they have disappeared without a trace. The possibilities can be studied. The examination of a series of well-preserved, younger rocks may disclose the time of appearance of
sterane hydrocarbons in the geological record. The study of a
number of ancient rocks with a varying thermal history may indicate if steranes have the intrinsic chemical stability to
persist for great lengths of time.

The organic matter in the Nonesuch Shale represents a unique
opportunity for Precambrian studies and forms a baseline with
which to compare results on older rocks that have not been so
well preserved. It is evident that a number of questions must
be answered before the molecular fossil concept can be extended
to earlier parts of the earth's history.

REFERENCES

Barghoorn, E. S., Meinschein, W. G., and Schopf, J. W., (1965). Paleobiology of a Precambrian shale, *Science 148*, 461-472.

Eglinton, G., Scott, P. M., Belksy, T., Burlingame, A. L., Richter, W., and Calvin, M. (1966). Occurrence of isoprenoid alkanes in a Precambrian sediment, *in* "Advances in Organic Geochemistry 1964" (Hobson, G. D., and Louis, M. C., eds.), pp. 41-71. Pergamon Oxford.

Fieser, L. F., and Fieser, M. (1959). "Steroids," VanNostrand-Reinhold, Princeton, New Jersey.

Giger, W., and Blumer, M. (1974). Polycyclic aromatic hydrocarbons in the environment: Isolation and characterization by chromotography, visible, ultraviolet and mass spectrometry, *Anal. Chem. 46*, 1663-1671.

Hoering, T. C. (1967). The organic geochemistry of Precambrian rocks, *in* "Researches in Geochemistry," Vol. 2 (Abelson, P.H., ed. pp. 89-111. Wiley, New York.

Hoering, T. C. (1971). Conversion of polar organic molecules in rock extracts to saturated hydrocarbons, *in* "Carnegie institution Year Book 70". pp. 251-256. Carnegie Institution of Washington, Washington, D.C.

Hoering, T. C. (1977). Olefinic hydrocarbons in the Bradford Pennsylvania crude oil, Chem. Geol.

Hunt, J. M., and Forsman, J. P., (1958). Insoluble organic Matter (Kerogen) in sedimentary rocks of marine origin, *in* "Habitat of Oil" (Weeks, L.G., ed. pp. 747-778. American Association of Petroleum Geologists, Tulsa.

McKirdy, D. M. (1974). Organic geochemistry in Precambrian research, *Precambrian Research 1*, 75-137.

White, W. S., and Wright, J. C. (1966). Sulfide mineral zoning in the basal Nonesuch Shale, Northern Michigan, *Econ. Geo.. 61*, 1171-1190.

16

MICROORGANISMS PRESERVED
IN CHERTY STROMATOLITIC DOLOMITE
OF THE MWASHYA GROUP FROM N'GUBA,
MULUNGWISHI, AND SHITURU LOCALITIES,
SHABA, ZAIRE, CENTRAL AFRICA

Anna-Stina Edhorn

Department of Geological Sciences
Brock University, St. Catherines, Ontario, Canada

Stromatolites and algal mats occur in some horizons in late Precambrian rocks of the Mwashya Group in the Roan Series of the Katanga System in several localities in Shaba, Zaire, Central Africa.

Samples from such horizons studied in thin sections from N'Guba, Mulungwishi, and Shituru localities contain well-preserved microorganisms. Spherical, yellowish to red to black cells of varying sizes, single and in accumulations of different configurations, are dominant in the Shituru and Mulungwishi and also occur frequently in the N'Guba material together with larger sporelike cells, some probably representing spore capsules. Also present are colonial coccoid algae. Filamentous algae showing spiraled inner structures appear in some of the Mulungwishi material and probably represent a more advanced algal group such as the green algae. Tubelike filaments mainly without inner structure and with distinctive circular openings and cross sections are also present.

Similar accumulations of spherical cells as in the Mwashya rocks have been encountered in the Upper cherty stromatolitic horizons of the Gunflint Formation, Animikle Series, Thunder Bay, Ontario, Canada.

Some comparisons between these findings will be attempted.

INTRODUCTION

The discovery of stromatolites and algal mats in rocks of the Katanga System in Shaba, Zaire, Central Africa, adds new interesting aspects involving microorganisms and the influence their activities might have on rock formation in general and mineralization in particular since the stromatolites and the algal mats occur in close association with important copper mineralization.

J. Lefebvre of the Union Miniere Explorations and Mining Corporation Limited made it possible for me to study some of their thin sections of the Katanga System, which includes the well-known copperbelt of Central Africa. The samples were collected from the Mwashya Group in N'Guba, Mulungwishi, and Shituru localities, Shaba, Zaire.

The association of the stromatolites and a certain mineralization is an interesting one, but the aim in this chapter is to report on the microorganisms found in the thin sections provided from some of the algal horizons.

Because of the economic value of the Copperbelt and the association of the stromatolites and algal mats within the mineralized zone, much interest have been focused on the presence of the stromatolites and other algal structures in the sequence. Hence the possible influence in respect to the mineralization that can be attributed to organisms or actions by organisms that built stromatolites or algal mats has been the subject of considerable speculations.

Pioneer papers in the subject are by Ashley (1937), Jamotte (1941, 1944), Cahen and Mortelmans (1941), and Hacquaert (1943), followed by papers in recent years by Garlick (1961, 1964), Malan (1964), and Paltridge (1968), to mention a few. Important stratigraphic work has been done by Oosterbosch (1962), Francois (1973, 1974), Lefebvre (1973, 1974), and Cahen (1974).

The types of microfossils that could be present in the stromatolitic beds have apparently not been studied to any great extent. According to Binda (1972), who has reported some spheromorphs from Zambia, not very much has been reported regarding microfossil content from the stromatolitic horizons in the Copperbelt. Cahen et al. (1946) have reported some findings, as well as Hacquaert (1943), who described some Girvanella in the northern equivalent group to the Mwashya, the Kitondwe Group (Dumont, 1971).

GEOLOGIC SETTING

The Katanga System, so named by Van Doorninck (1928) and Mendelsohn (1961), has a thickness of 10,000 m in Shaba and is traditionally subdivided in lithostratigraphic units. The upper part or Kundelungu Supergroup (about 900-600 m.y.) consists of red beds, tidal deposits, and probably tillites. The Lower part or Roan Supergroup (about 1300-900 m.y.) is formed of a cyclic succession of red beds and tidal deposits, some of them associated with rich copper and cobalt stratiform mineralization. The upper cycle represents the red beds of the Shituru Group, underlying the Mwashya Group.

The oldest beds of the Mwashya Group (970 m.y.) consist of back reef stratified dolomites (algal mats) intermixed with basic pyroclastic rocks and jasper and hematite beds. Above this we find a reef facies with stromatolites, oolites, and pisolites, followed by felsic tuff-lavas, jasper, and in some cases massive pyrite. Finally the group ends with fore reef sediments; like dolomitic sandstones and black shales. In some places the Mwashya Group contains some syngenetic or diagenetic copper sulfides (chalcopyrite and bornite) in the algal dolomite of the back reef.

The lowest cycle of the Roan Supergroup is the red beds and stromatolitic dolomitic shales of the Mines Group, containing the most important mineralization of Shaba. In the Zairian Copper district, the metamorphism has never been higher than the sericite-chlorite phase. The three localities, N'Guba, Mulungwishi, and Shituru are all situated within a 30-km radius.

N'GUBA LOCALITY

In the N'Guba locality the thickness of the Mwashya Group is about 105 m. The sections containing preserved microfossils are beds of reddish brown dolomite with oolites and stromatolites of Collenia type.

Yellowish cells with accicular surfaces are encountered in the above-mentioned beds and in oolitic dolomite adjacent to the stromatolite beds. The cells have a size range of 2 to 4 µm in apparently young cells and 9 to 11 µm in the mature organisms, whose cell walls often are split open (Figs. 1 and 2). Some of these cells are 15 to 17 µm in diameter. Accumulations of tightly packed, small spherical cells closely associated with the larger ones probably represent reproduction stages (Figs. 4 and 5). Some polygonal cells about 30 to 40 µm in diameter and with a slit and some with "spores" dispersed around that slit indicate that the polygonal cells most likely represent spore-capsules (Fig. 3).

Also present are round cells enclosed in rings of laminae. They consist of small cells ranging in size from 15 to 23 µm and larger ones from 60 to 78 µm. They form densely packed mats where cell division and mutual compression cause variations in shapes and sizes. It seems most likely that they represent colonial coccoid algae, similar to recent green algae *Glöeocystis* (Figs. 6-8) and reminiscent of the *Cumulosphaera* Edhorn from the Gunflint Formation, which are preserved in a less recrystallized chert carbonate matrix (Figure 9 and 10). They might be compared with *Bigeminococcus* Schopf-Blacic, but the

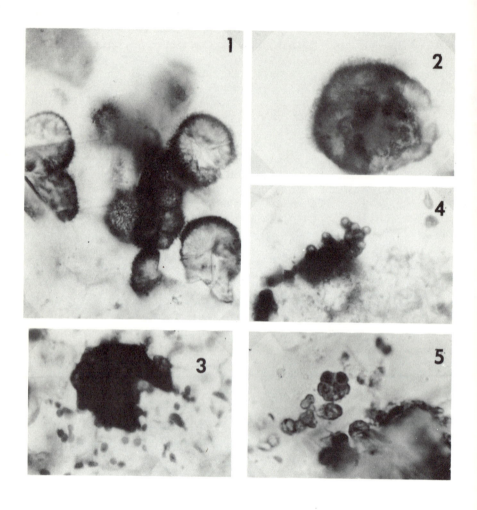

FIGURE 1. Accicular cells split open, from N'Guba locality. x 1040.

FIGURE 2. Accicular cell with inner structure, from N'Guba locality. x 1300.

FIGURE 3. "Spore capsule" discharging spores. x700.

FIGURE 4. Reproduction stages in coccoid algae. x 700.

FIGURE 5. Reproduction stages in coccoid algae. x 1750.

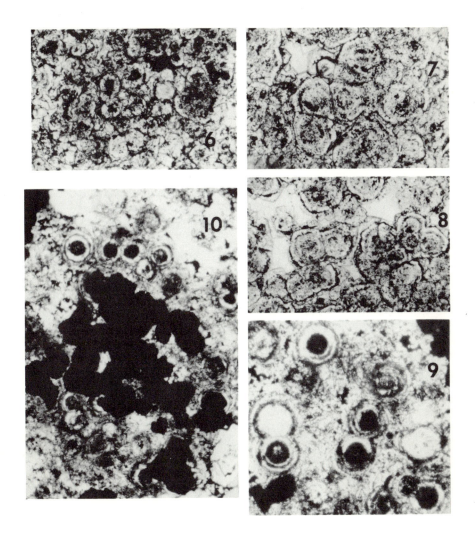

FIGURES 6,7.   Cumulosphaera Edhorn, from N'Guba locality
x 140.
   FIGURE 8.   Cumulosphaera Edhorn, from N'Guba locality. x 110.
   FIGURE 9.   Cumulosphaera Edhorn, from the Gunflint Formation, Canada. x 700.
   FIGURE 10.   Cumulosphaera Edhorn, from the Gunflint Formation, Canada. x 450.

size differences are large and the individual cells are not enclosed in individual sheaths in the *Bigeminococcus*.

An interesting feature observed in the *Cumulosphaera* Edhorn, is the mineralization of pyrite in the cell core and also
sometimes enclosing the cell sheath, hence covering the whole
cell structure or several of them together, only preserving the
circular shape of the cells (Fig. 10). Such mineralization
exists also in the Mwashya Group (Fig. 6). Similar observations
of cells wholly enclosed in pyrite have been made and are described by Love (1958) and Schopf and Blacic (1971).

SHITURU LOCALITY

Reddish brown cells, psilate or sculptured, occur in the
cherty silty beds in the Shituru locality. They form small
accumulations of different configurations, clusters and chains,
pairs and tetrads, but also occur as individuals (Figs. 14 and
15). They are quite abundant along bands of disintegrating organic material (Fig. 13). They have quite a wide size
range, from a few micrometers to 9-20 µm. Individual cells
can be spherical, pyriform, or angular. Similar extant coccoid
blue green algae are, for example, *Xenococcus* Thuret, which are
epiphytic on filamentous algae. They form endospores and also
multiply rapidly by fission.

The Shituru cells, with their large size range and habit of
accumulating into distinctive configurations, show a close similarity to spherical cells found in the Gunflint Formation and
interpreted by Edhorn (1973) as probably belonging to the green
algae *Sphaerocystis*. The occurrence of such spherical cells
along bands of filamentous algae laminations has lately been
recognized by the author in some Gunflint stromatolitic material. In such cases, as in the Mwashya rocks, the affinity of
these spherical cells points toward a *Xenococcus* type alga
(Figs. 14 and 15).

MULUNGWISHI LOCALITY

In a blackish, oolitic dolomite in the Mulungwishi locality,
preserved filaments are encountered, some of them showing a
spiraled inner structure, which could possibly be interpreted
as preserved chloroplast as in the green algae of spirogyra
type. However, there is an equal possibility that they represent the blue-green algae of oscillatorian type. They have
been preserved in geo-chemically altered portions of the dolomite

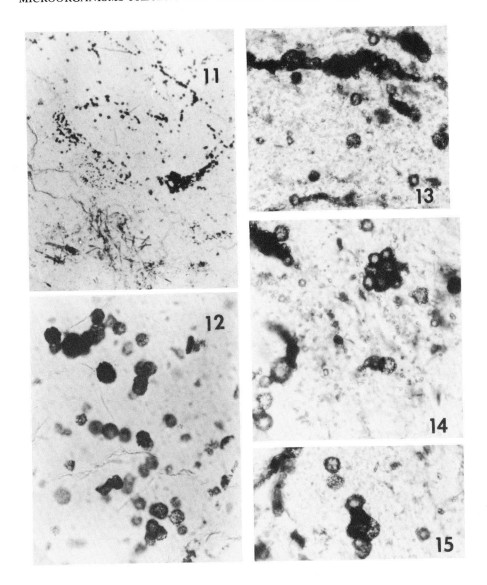

FIGURE 11. Accummulation of spherical cells from the Gunflint Formation. x 175.

FIGURE 12. Spherical cells from the Gunflint Formation, enlarged. Note pyritization of cells. x 700.

FIGURE 13. Spherical cells along laminae, from Shituru locality. x 560.

FIGURE 14, 15. Spherical cells of different sizes in pairs and cluster, from Shituru locality. x 700.

in dark bands and patches. The width of these filaments is
about 2 µm (Figs. 16 and 17).

Other filaments mainly without inner structure and with
tubelike shape are also present. The width of those tubes is
2-3 µm. The often noncollapsed tubes have circular to oval
openings or cross sections, which are easily recognized. They
occur as short tubes sometimes bent to show a circular opening
in each end of the tube and sometimes U-shaped, but mostly forming intricate, intertwisted bundles (Fig. 18). Long, almost
straight tubes are sometimes present (Fig. 19). Only in very
limited cases can constrictions be observed in the walls, de-
noting probable segmentation.

The rigidity of the filamentous tubes, the circular openings, and the formation of intricate, intertwisted bundles are
very characteristic for *Girvanella* (Johnson, 1966; Wray 1969).
This controversial alga is very common already in the early Cambrian limestones and can therefore be expected to be present in
the late Precambrian as well. Hacquaert (1943) has reported
some *Girvanella* in the northern equivalent group to the Mwashya,
the Kitondwe Group, and even though the illustration is poor in
the report, one can recognize a few tubes, which seem to fit
in together with Hacquaert's description to a general picture
of *Girvanella*.

The appearance of *Girvanella* in the Proterozoic might have
some interesting implications, because *Girvanella* evidently became a very dominant alga and there are many indications that
the filamentous, stromatolite-forming algae were forced out from
an earlier dominance by these resistant, cohesive algae.

In the Cambrian, *Girvanella* flourished and is considered to
have been a limestone builder. At the same time, the formation
of stromatolites decreased and it appears there was a relation
between these events.

Masses of yellow, reticulate cells are surprisingly well
preserved in the same type of bands of alteration as the filaments. They are spherical, but often squeezed to an ovoid
shape about 6-9 µm in diameter.

Some hyaline, psilate cells about 6 µm in diameter (Fig. 20)
could be compared with *Chlamydomonopsis* Edhorn from the Gunflint.
They represent the same type of cells with intracellular bodies
or organelles as outlined by Schopf (1974) from the Bitter
Springs biota.

DISCUSSION

It was interesting to postulate the similarities between
the much older Gunflint organisms preserved in chert and the
organisms preserved in cherty dolomites of the much younger
Mwashia rocks. That proves the flexibility and tolerance and

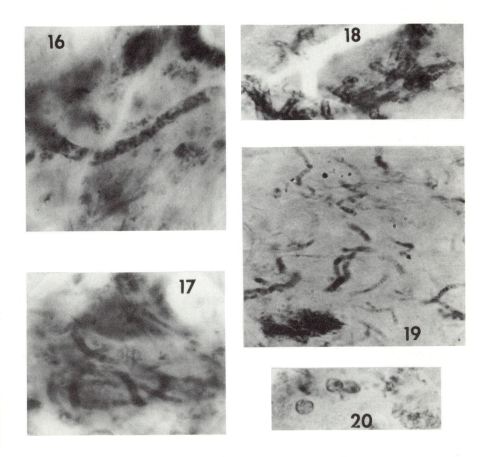

FIGURE 16.   Spiraled inner structure in filament, from
Mulungwishi locality.   x1235.
FIGURE 17.   Spiraled inner structure in filaments, from
Mulungwishi locality. x650.
FIGURE 18.   Girvanella tubes, from Mulungwishi locality.
x650.
FIGURE 19.   Girvanella tubes showing segmentation and/or
inner structure, from Mulungwishi locality. x650.
FIGURE 20.   Chlamydomonopsis Edhorn, Mulungwishi locality.
x650.

adaptability of these organisms in a changing biosphere. This ability has secured their survival up to recent time. The discovery of *Girvanella* tubes in the 970 m.y. old rocks of the Mwashya Group might add some explanation to the speculation of the drastic decline of stromatolites before any known grazers had appeared and would support Monty's (1972) suggestion that competition played an important role among the early algae population. As soon as a strong competitor evolved, the balance that had been achieved between the existing stromatolite-forming algae was broken and the special conditions that favored such a stabilization for those algae no longer exist. *Girvanella* was apparently such a strong competitor that the stromatolite layers disappeared abruptly where *Girvanella* came in, as found in lower Cambrian rocks (Edhorn and Anderson, 1976).

Stanley's (1976) comparison with today's conditions, where areas regarded as favorable for algal growth are barren and thus imply cropping activity as the only cause of algal elimination to contradict Monty's (1972) claim   eliminating effect of one algal population through strong competition by another, cannot really be representative, so many factors other than cropping and algal competition could prevent algal growth (not the least of which is the action of man, who could also reverse the situation). However, the combined eliminating effect by cropping and competition was and is probably the most common.

The presence of *Girvanella* and colonial, coccoid algae in some horizons of the Copperbelt also creates new imputs of speculations regarding the influence of these algae concerning the accumulation of mineral matter. The mineralization of pyrite in and around the cells and colonies of the coccoid algae, as illustrated both in the Gunflint and Mwashya rocks, gives evidence that such influence is real. Tough, gelatinous algal mats certainly can be the breeding place of many matters.

The blue-green algae and apparently bacteria were among the first organisms to evolve. They are still with us and most likely will be the last to disappear. If the same development occurred on Mars as on the earth, remains of these algae will be preserved in the rocks and soil and will probably supply Viking with the crucial organic compounds we are all waiting for to postulate that life exists or has existed on Mars.

ACKNOWLEDGMENTS

This research project was made possible by the opportunity
kindly given the author by Union Miniere Explorations and
Mining Corporation Limited to study their thin sections from the
Katanga System. Thin sections, literature, and maps were put at
my disposal, for which I am very grateful. Special thanks go
to J. J. Lefebvre, who initiated the study, for valuable information and interesting discussions, and to the Department of
Geological Sciences, Brook University, for the use of their
excellent equipment and facilities.

REFERENCES

Ashley, B. E. (1937). *J. Geol. 45*, 332.

Binda, P. L. (1972). *Geol. En Mijnbouw 51*, 315.

Cahen, L., and Mortelmans, G. (1941). *Com. Spec. du Katanga.*

Cahen, L., Jamotte, A., and Mortelmans, G. (1946). *Ann. Soc.
Geol. Belg. 70*, B55.

Cahen, L. (1974). In *"Gisements Stratiformes et Provinces
Cupriféres" (P. Bartholmé, ed.), p. 57. Liège.*

Dumont, P. (1971). Revision générale du Katangien. Unpublished
thesis. Free University of Brussels, Belgium.

Edhorn, A. S. (1973). *Geol. Assoc. Can. Proc. 25*, 37.

Edhorn, A. S. and Anderson M. M. (1976). Algal remains in the
Lower Cambrian Bonavista Formation, Conc. Bay, S. E. New-
foundland (in press).

Francois, A. (1973). Etude Géologique, Gecamines, Likasi,
Shaba Zaire, p. 120.

Francois, A. (1974). In "Gisements Stratiformes et Provinces
Cupriféres" p. 79 Liege.

Garlick, W. G. (1961). In The Geology of the Northern Rhodesian
Copperbelt (F. Mendelsohn, ed). p. 146. Macdonald, London.

Garlick, W. G. (1964). *Econ. Geol. 59*, 416.

Hacquaert, A. L. (1943). *Natuurwetenschap. Tijdschrift 25*, 33.

Jamotte, A. (1941). *Com. Spec. Du Kantanga 6.*

Jamotte, A. (1944). *Com. Spec. du Katanga 22.*

Johnson, J. H. (1966). *Colorado School of Mines, Quart. 61* (1), 162.

Lefébvre, J. J. (1973). *Extrait Ann. Soc. Geol. Belgique. 96*,
197.

Lefébvre, J. J. (1974). In *Gisements stratiformes et Provinces
Cupriféres"* (P. Bartholome, ed.). p. 103. Liege.

Love, L. G. (1958). *Quart. J. Geol. Soc. London 113*, 429.

Malan, S. P. (1964). *Econ. Geol. 59*, 397.

Mendelsohn, F., ed. (1961). "The Geology of the Northern Rhodesian Copperbelt", p. 523. Macdonald, London.

Monty, C. (1972). *Geol. Rundsch. 61*, 742.

Oosterbosch, R. (1962). *Gisements stratiformes de cuivre en Afrique, Symp. 1, Ass. Serv. Geol. Afr. (Copenhagen, 1960)*, p. 71.

Paltridge, J. M. (1968). *Econ. Geol. 63*, 207.

Schopf, J. W. and Blacic, J. M. (1971). *J. Paleontol. 45*, 925.

Schopf, J. W. (1974). *Origins of Life 5*, 119.

Stanley, S. M. (1976). *Paleobiol. 2 (3)*, 209.

Van Doorninck, N. H. (1928). De Lufilische Plooiing, G. Naeff's Gravenhage, Pp. 1-201.

Wray, John, L. (1969). *Proc. N. Am. Paleontol. Convent. Part J*, 1358 (publ. 1971).

INDEX

A
B
C   8
D   9
E   0
F   1
G   2
H   3
I   4
J   5